Hamlyn all-col

Alan Isaacs and
Valerie Pitt

Physics

illustrated by Whitecroft Designs

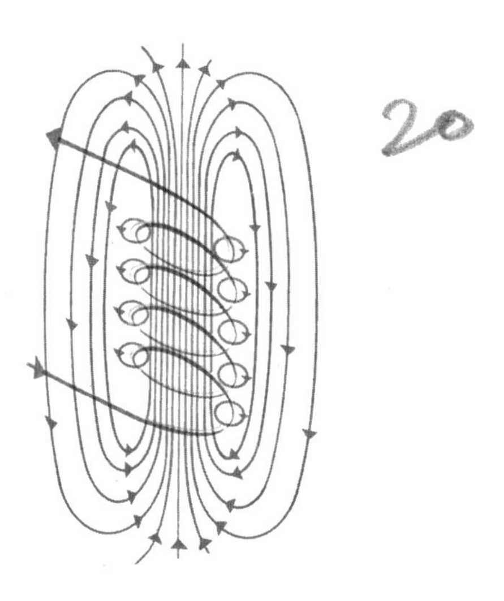

Hamlyn - London
Sun Books - Melbourne

FOREWORD

The aim of this book has been to give an account of what physics is about, rather than to furnish a comprehensive, didactic text. While covering the classical divisions of the subject the central theme of the book is energy. Much attention is therefore devoted to the interrelationships between different forms of energy and to the production and use of energy in modern life.

Physics is very much a science of measurement; it is therefore essential to include some equations even in a book of this type. These should not, however, prove too daunting to the non-mathematical reader.

M.T.B.

Published by the Hamlyn Publishing Group Limited
London · New York · Sydney · Toronto
Hamlyn House, Feltham, Middlesex, England
In association with Sun Books Pty Ltd., Melbourne

ISBN 0 600 36983 8
Photoset by BAS Printers Limited, Wallop, Hampshire
Colour separations by Schwitter Limited, Zurich
Printed in Holland by Smeets, Weert

CONTENTS

The velocity of the spear, V_R, is the resultant of the log's velocity, V_B, and the velocity, V_S, with which the spear leaves the hunter.

INTRODUCTION – WHAT IS PHYSICS?

In order to survive, primitive man had to defend himself against and compete with a great variety of fierce and hungry animals. A number of factors combined to enable him to become the first user of tools and weapons: perhaps the most important was his upright stance which enabled him to develop the use of his hands. Manual dexterity required a larger brain, and the larger brain not only made him better at defending himself and hunting but also stimulated his natural curiosity. He began to ask 'how' and 'why', to try to relate cause and effect.

Through ignorance and fear he would, at first, have attributed many *natural* effects to *supernatural* causes, but when he began to notice that certain effects always followed a particular cause, the systematic study of nature called science was born. Modern science is a vast field of human activity that concerns itself with both the living and the inanimate world; the branch called *physics* deals with the behaviour of matter and its relation to energy; it is therefore fundamental to all the sciences.

Perhaps the most important thing that the first hunters had to understand was motion. To throw a rock or primitive spear at a moving target, or when moving oneself, requires an awareness of two fundamental physical quantities – distance and time. Distance divided by time is speed, and relative speeds are particularly important in hunting.

In physics, the word velocity is used to describe a speed in a particular direction. Speed is called a *scalar* quantity because it has only magnitude; velocity is called a *vector* because it has both magnitude and direction.

Motion remained vague and instinctive until the Greeks made a serious attempt to understand it. Aristotle concluded, wrongly as it turned out, that a force was necessary in order to explain motion. It was Newton (and Galileo) in the seventeenth century who discovered that it is the forces that oppose motion that require an explanation rather than the motion itself. A spacecraft will glide through space indefinitely once it has been set in motion because there are no forces to oppose it.

Velocity remains constant in the absence of forces. Thrust accelerates the rocket; frictional and wind forces decelerate the cone until it comes to rest.

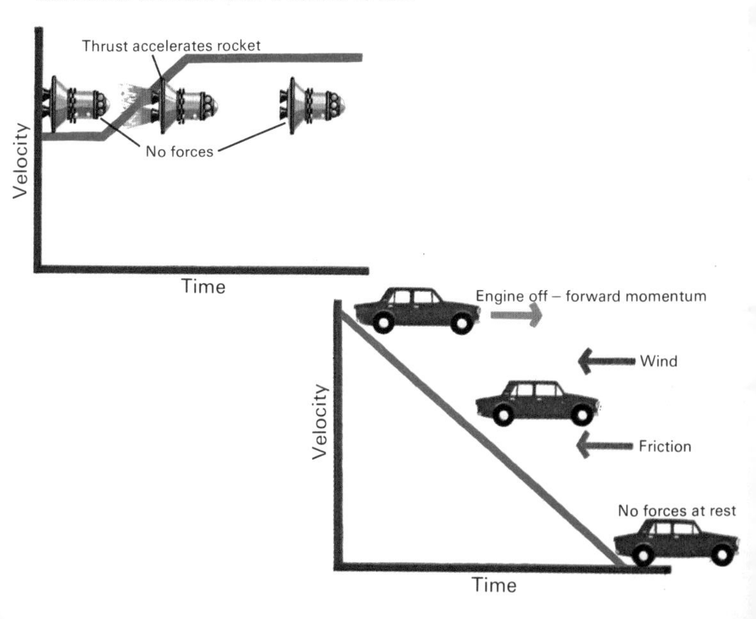

MECHANICS

The velocity-time graphs on the previous page provide a means of relating velocity and acceleration with time and distance covered. If the initial velocity of the body is u and the final velocity after time t is v, then the gradient of the graph is its acceleration, a, or rate of change of velocity, $(v - u)/t$. The area under the graph represents the distance, s, covered by the body in the time t.

These notions are only concerned with rates of motion, they do not take *forces* into account. The relationships between acceleration and force follow from Newton's discovery that bodies continue to move at constant velocity unless acted upon by a force: this is Newton's *First Law of Motion*.

Newton's *Second Law of Motion* makes clear exactly what is meant by a force. It says that the rate at which the velocity of a body changes (its acceleration or deceleration, a) is proportional to the force, F, acting on it. The constant of proportionality is the mass, m, of the body. In symbols, $F = ma$. In the modern system of units (known as SI), the unit of force, called the *newton,* is so chosen that 1 newton (N) gives a mass of 1 kilogram (kg) an acceleration of 1 metre per second per second (m/s^2).

The *mass* of a body is not the same as its *weight*. Weight is a force – the force with which the earth's gravity attracts the body. Therefore, according to Newton's second law, the weight of the body is equal to its mass multiplied by the acceleration, g, due to gravity. That is $W = mg$. Unfortunately the words mass and weight are used interchangeably in everyday speech. In physics however, it is essential to make the distinction: mass is measured in kilograms and weight in newtons.

Newton's *Third Law of Motion* states that for every force there is an equal force opposing it. If you lay a book on a table, there is the downward force of the book's weight on the table. If this was the only force on the book it would move downwards – it does not do so because there is an equal reaction in the opposite direction. In the same way, the combustion gases of a rocket produce a downward thrust; the reaction to this thrust on the rocket pushes it upwards.

Velocity-time graph of accelerating body. Acceleration $a = AC/BC = (v - u)/t$, hence $v = u + at$.
In time, t, the distance travelled, s, is given by the area under the graph – $s = ut + \frac{1}{2}at^2$.
Combining these equations – $s = \frac{1}{2}(u + v)t$ and $v^2 = u^2 + 2as$.

From Newton's Third Law of Motion the time of burn, t, required for the rocket to reach escape velocity, v_e, can be calculated. Reaction thrust $= R$, downward force (weight of rocket) $= mg$; hence $R - mg = ma$. Since $v_e = at$, $t = mv_e/(R - mg)$. (This ignores any change in the rocket's mass as fuel is used up)

Gravitation

Weight is the name given to the force on a body that results from the earth's gravitational attraction. But what is this gravitational attraction? Newton discovered that between all bodies having mass there exists an attractive force: this force is one of the four fundamental types of force in the universe. The others are the electromagnetic force and the two types of nuclear force. Although Newton did not know the cause of the gravitational force (it has been partially explained by Einstein), he did discover that its strength is proportional to the product of the interacting masses and inversely proportional to the square of the distance, s, between them. This is known as Newton's Law of Gravitation; in symbols: $F = Gm_1m_2/s^2$, where m_1 and m_2 are the masses and G is the *Gravitational Constant* (which has the same value everywhere in the universe).

If M is the mass of the earth and S is its radius, the force of attraction, F, on a mass m at the earth's surface is found from this law to be GMm/S^2. F is equal to the weight, w, of the body, i.e. $F = w = mg$, where g is the acceleration due to gravity (see page 6). Comparing these two equations, it follows that $g = GM/S^2$. Inserting the known values of these constants gives $g = 9.81$ m/s^2: this value varies slightly in different localities due to variations in the distance from the earth's centre of mass. Using values of m and s appropriate to the moon gives a lunar acceleration due to gravity of $g/6$. A lunar explorer has the same mass as he has on earth, but a sixth of the weight – it is this that enables him to take such giant leaps.

If a body of mass m moves in a circle radius r with a uniform speed v, there is an acceleration towards the centre of the circle because even though the speed (scalar) does not change, the velocity (vector) does – the body is constantly changing direction and therefore velocity. This *centripetal* acceleration, as it is called, is equal to v^2/r, and therefore the force towards the centre of the circle is, according to Newton's second law, ma or mv^2/r. The equal and opposite reaction to the centripetal force is called the *centrifugal* force.

As the moon circles round the earth it experiences two forces – one is the gravitational attraction to the earth, the other is the centrifugal force. Because the moon stays in orbit, it follows that these forces must be equal, that is $GMm/d^2 = mv^2/d$, where m is the mass of the moon and d the distance between the centre of the earth and the centre of the moon. Substituting known values for G and M, the velocity of the moon round the earth, $v = 2 \times 10^7/\sqrt{d}$ m/s. This approximate formula (it does not take account of other bodies in the solar system and assumes circular orbits) not only applies to the moon, but also to any other earth satellite: it therefore enables the height of any satellite in stable orbit to be related to its velocity.

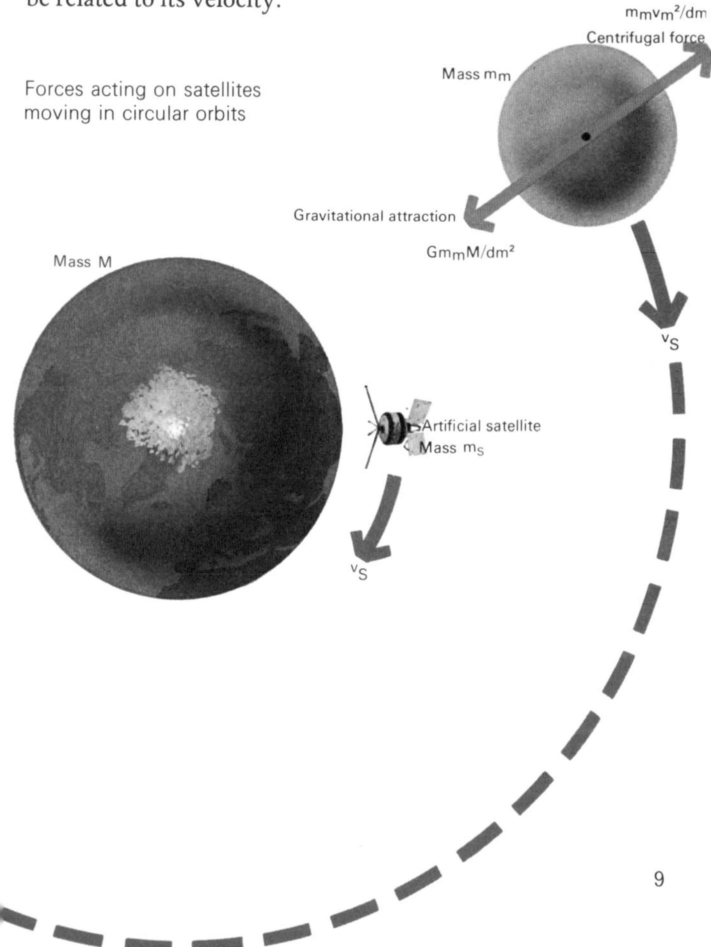

Forces acting on satellites moving in circular orbits

Work and energy

Work and energy are words that are used rather loosely in everyday speech; in physics they have very precise meanings.

In the illustration the man looks as if he is doing a great deal of work, but in physical terms he has done no work until the boulder moves. In physics, *work done, w,* is equal to the force, *F,* multiplied by the distance moved, *s,* in the direction of the force: $w = Fs$. If the force is in newtons and the distance in metres, the work is in newton metres or *joules* (the joule – the SI unit of work or energy – is defined as 1 newton metre).

The use of the word *energy* was only finally settled in the nineteenth century. It is now recognized as one of the principal concepts of physics: it is simply the ability of a body or system to do work (work, of course, in the physicist's sense).

When the man has managed to move the boulder to the edge of the cliff (a distance *d*) the work done is Fd. The boulder is now able to work – it has energy. But the energy is potential as long as it is at rest. The *potential energy* of the boulder is the work it would do by falling as far as it could under the earth's gravitational force. If the mass of the boulder is *m*, the gravitational force on it is mg, and if the height of the cliff is *h*, the potential energy of the boulder is mgh.

The pusher's efforts are finally rewarded and the boulder is on its way down. When it has fallen 30 per cent of the way it has lost 30 per cent of its potential energy. This amount of energy has not disappeared – energy never disappears – it has been converted into energy of motion, or *kinetic energy* as it is called. The kinetic energy of a body is proportional to the square of its velocity. This is because kinetic energy equals Fs, and as $v^2 = 2as$, $Fs = mav^2/2a = \frac{1}{2}mv^2$.

After the body has fallen 30 per cent of the way $\frac{1}{2}mv^2 = 0.3\ mgh$ ($s = 0.3h$). The velocity is then $\sqrt{0.6\ gh}$ – it does not depend on the mass of the body. This is only true in a vacuum; a feather falls more slowly than a brick because of a different air resistance.

In the bottom picture, the boulder has come to rest at the foot of the cliff. It now has neither potential nor kinetic energy: all its energy has been dissipated in making a hole in the ground, heating up the earth and air, and making a noise on impact. Energy can be converted from one form to another, but can neither be created or destroyed – this is the principle of *energy conservation*. However, since Einstein discovered that mass and energy can be converted into each other

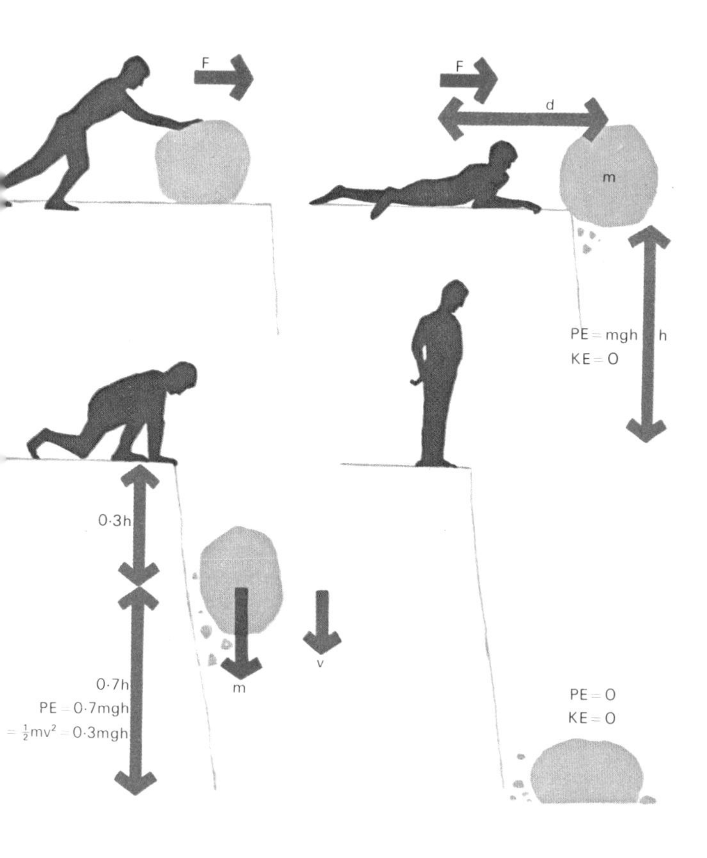

Work, potential energy and kinetic energy

according to his equation $E=mc^2$ (where c is the velocity of light) the conservation of energy only holds for systems in which there is no change of mass.

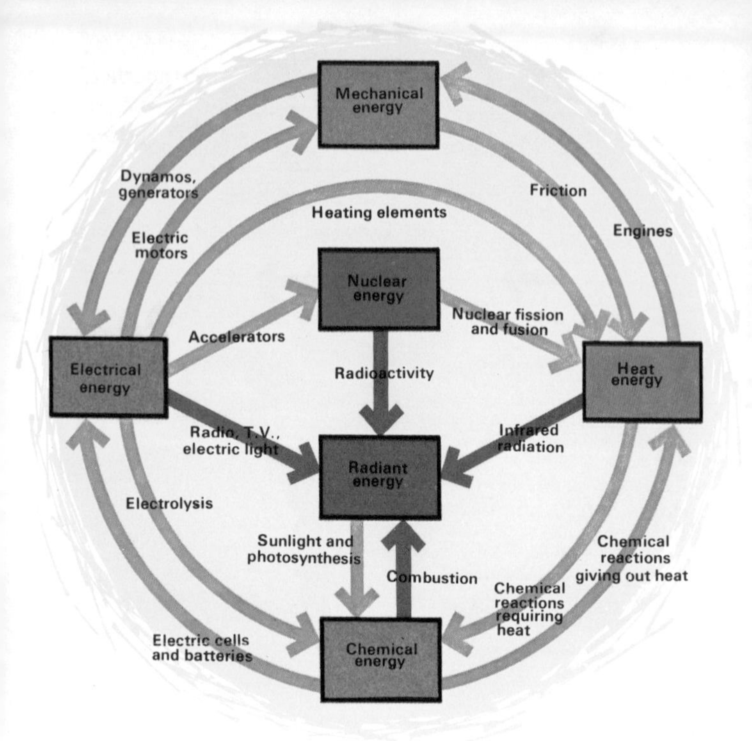

Interconversion of different forms of energy

Forms of energy

The total of the potential energy and the kinetic energy of a body or system is sometimes called its *mechanical energy* – this is the energy that can do work by means of some kind of machine. But energy can also be used or stored in a variety of other ways and it is often convenient to distinguish between less fundamental forms of energy than potential and kinetic. *Chemical energy*, for example, is the energy stored in the atoms and molecules of a substance. When chemical reactions take place, some of this chemical energy may be converted to *heat energy* – as when a fossil fuel (coal, oil, natural gas) is burned – and during combustion some of the heat may be given off as *radiant energy* in the form of *infrared radiation*.

A heat engine is a device for converting heat energy into *mechanical energy*; but when one surface rubs against another some of the mechanical energy is converted back into heat by friction.

Electricity is another form of energy – it can be produced from chemical energy in an electric cell or battery, or it can be produced from mechanical energy in a generator or dynamo. This *electrical energy* can be converted into radiant energy in the form of radio waves, light, ultraviolet radiation, or X-rays. It can also be converted back into mechanical energy by an electric motor, which is the reverse of a generator.

The ultimate energy of matter is locked within atomic nuclei. The immense forces that hold the particles together in a nucleus can be released to produce heat and radiant energy in a nuclear reactor (or bomb).

These are some of the energy transformations illustrated. Each type of transformation can only occur in the presence of matter. Energy can exist in empty space, but only in the form of radiant energy.

All our energy comes to us in the form of radiant energy from the sun, except for the radioactive energy of the earth's rocks. Comparison of the energy reserves of the various fossil fuels with nuclear sources shows the importance of developing a method of producing energy by nuclear fusion.

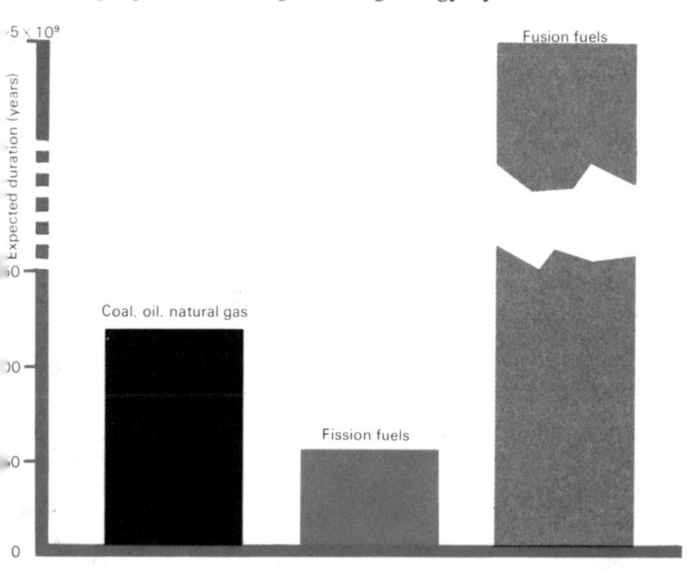

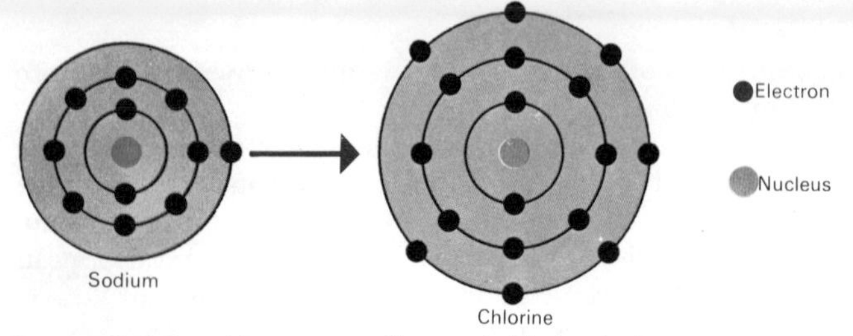

Electrovalent bond between sodium and chlorine to form common salt. (*Opposite*) Covalent bond between oxygen and hydrogen to form water, and the chemical reaction between oxygen and hydrogen releasing energy.

CHEMICAL ENERGY

All the matter in the universe is made from combinations of some hundred different kinds of atom. Each different kind of atom is called an *element*. Atoms consist of a central nucleus containing a fixed number of positively charged particles called *protons* and a fixed number of electrically neutral particles called *neutrons*. Negatively charged particles called *electrons* orbit around the nucleus, the number of electrons equalling the number of protons so that the whole atom is electrically neutral. These electrons have different energies and one way of depicting their different energy levels is to think of the electrons occupying a concentric series of shells (although in fact the orbits of the electrons are elliptical), each higher energy level being represented by a shell a little further from the nucleus.

Electron configuration of elements at the beginning of the periodic table with their mass numbers

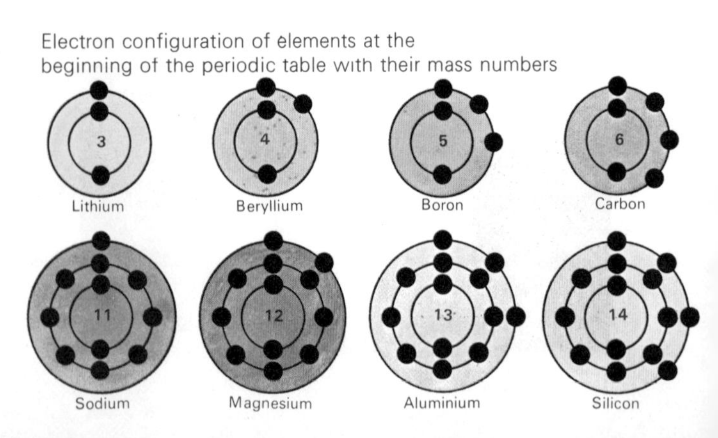

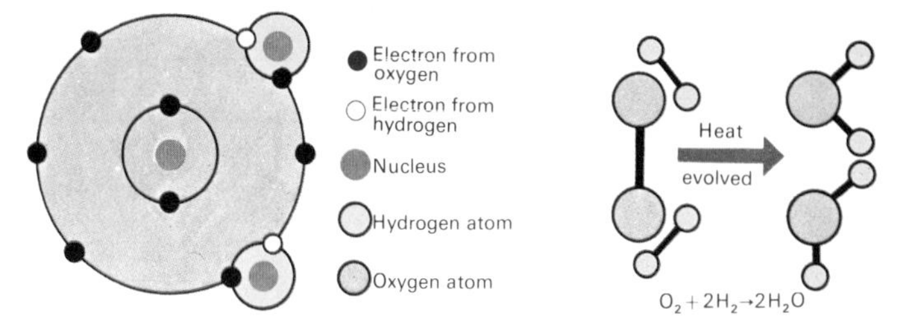

There are several rules concerning the number of electrons that can be contained in a shell. The most important is that atoms are unstable unless their outer shells contain eight electrons, except for hydrogen and helium which have only one shell that can contain only two electrons. Although there are only about 100 different elements, there are millions of chemically distinguishable substances, all of which are formed by combinations of atoms seeking to increase their stability by completing their outer electron shells. Atoms combine to form the molecules of a compound in one of two ways, or a combination of both these ways. In the *electrovalent bond* one atom donates an electron to another, so that both end up with eight electrons in their outer shells. In the *covalent bond,* an electron from one atom is shared between two combining atoms so that both complete their outer shells.

The chemical bonds between atoms represent stored potential energy. This potential energy, with the kinetic energy of the molecules, is the *chemical energy* of a substance.

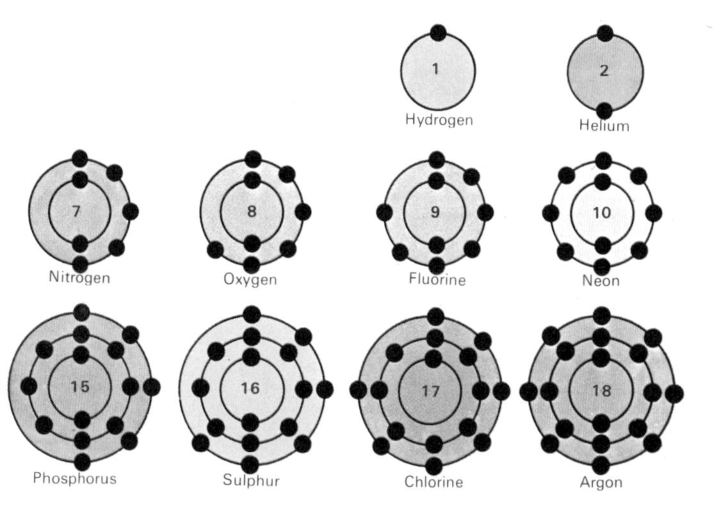

Gases

Matter can exist in three physical states: solid, liquid, or gaseous. In gases the atoms or molecules move about at random at enormous speeds, and as there are enormous numbers of them they are continually colliding with each other. An enclosed gas exerts its pressure as a result of molecular collisions with the walls of the container.

For a given quantity of gas, the smaller the space that it occupies the greater will be the number of collisions with the walls and hence the greater will be the pressure. The pressure, P, and the volume, V, are related by $P=k/V$, where k is a constant. This is known as *Boyles Law.*

The temperature of a gas is a measure of the *average* kinetic energy of its molecules. If the temperature is raised the average speed of the molecules is increased and there is therefore an increase of pressure if the volume is kept constant, or an increase of volume if the pressure is kept constant. For a rise of temperature of 1°C all gases expand by 1/273 of their volume at 0°C. This is known as *Charles' Law.*

This means that if the temperature is reduced to –273°C, the volume would theoretically become zero. This temperature is called the *absolute zero* (it is actually –273.15°C) and it is unattainable in practice. For many purposes, however, it is useful to have a temperature scale with zero at absolute zero. Temperature measured in this way, using a degree called the *kelvin,* is the *thermodynamic* or *absolute temperature.*

Boyle's Law relates pressure and volume at constant temperature.

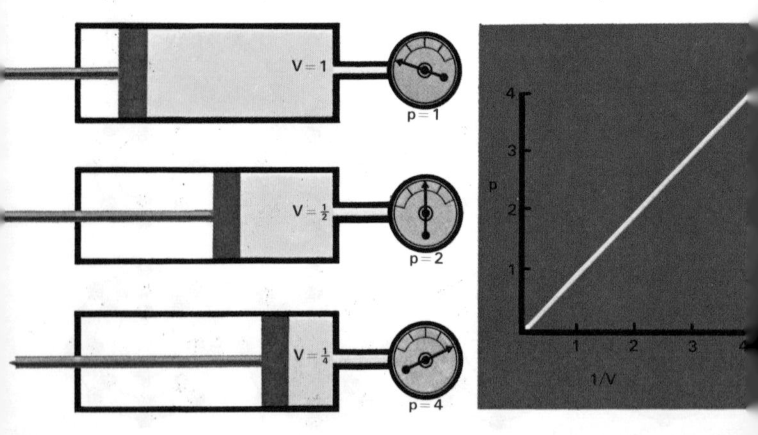

Boyle's and Charles' Laws can be combined into one equation, known as the *ideal gas law*: $PV=RT$. This applies to unit amount of gas, R being the *Gas Constant* and T the absolute temperature. Real gases deviate from this law as the pressure increases because as the molecules come closer together there are weak attractive forces between them.

Irrespective of the composition of a gas, equal volumes at the same temperature and pressure contain equal numbers of molecules (*Avogadro's Hypothesis*). At standard temperature and pressure there are 2.7×10^{19} molecules/cm^3 of gas and each molecule will experience some 5×10^9 collisions/second.

The volume of a gas is proportional to its temperature at constant pressure (Charles' Law) and its pressure is proportional to temperature at constant volume.

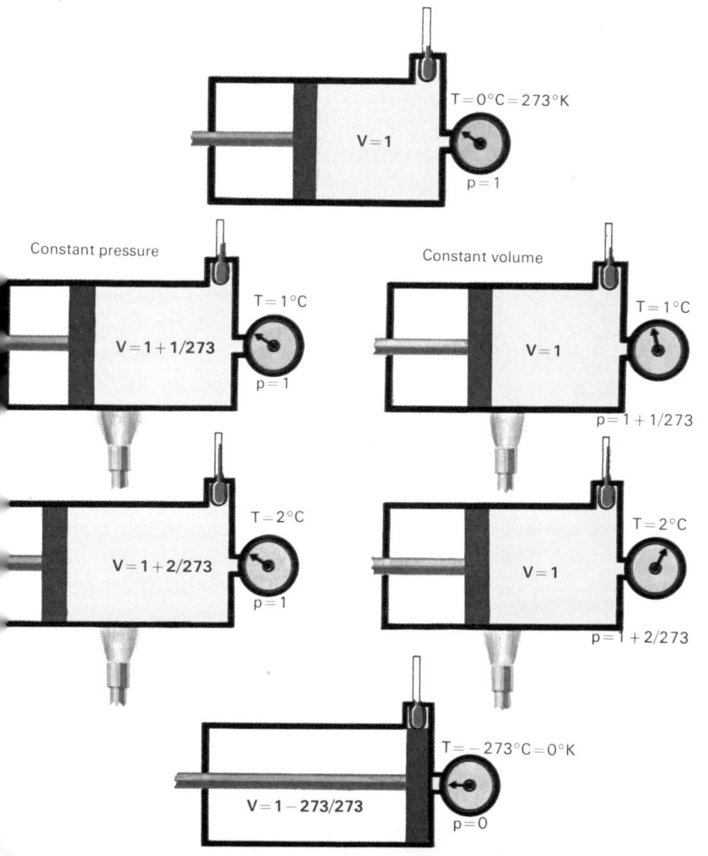

Liquids

When a gas is cooled, the average energy of its molecules decreases and at a certain temperature is no longer high enough to overcome the attractive forces that exist between molecules. The molecules tend to cluster together in groups under the influence of these forces, thereby greatly reducing the random motion of individual molecules. However, the groups are still free to slide over one another and adopt the shape of their container. The gas has become a *liquid*.

The molecules of a liquid are in continual motion and some of the more rapid ones on the surface will break free of the bonds holding them to their neighbours, causing a reduction of volume. This is called *evaporation*. The pressure exerted by the escaping molecules, which for a particular liquid at a given temperature will always be the same, is called the *vapour pressure* of the liquid. This increases with temperature and becomes equal to the atmospheric pressure at the *boiling point* of the liquid.

Water is the most common liquid on the earth's surface, but its structure has only recently been understood. As a gas (steam), it consists of triatomic molecules, but in the solid and liquid states additional forces, called *hydrogen bonds,* form between adjacent molecules. These hydrogen bonds, which are ten times weaker than covalent bonds, are electrical attractions between the positively charged nucleus of a hydrogen atom and the small negative charge around an oxygen atom in a neighbouring molecule. This small negative charge is a result of a tendency for electrons in a water molecule to cluster round the oxygen atom. In ice, a three dimensional tetrahedral structure builds up, each water molecule being surrounded by four neighbours in a very open structure. When ice melts, this open structure collapses, and the water molecules pack closer together. Water is therefore denser than ice and consequently ice floats on water. Even though the tetrahedral structure of ice breaks down on melting, small aggregates of molecules continue to survive up to 4°C; after this, density decreases in the normal way with temperature. Water therefore has the unusual property of a maximum density at 4°C. Hydrogen bonds continue to hold water molecules together until the water boils at 100°C.

Hydrogen bonding of water molecules forms an open lattice structure (*upper*). Fish survive below ice which floats on the water's surface (*lower*).

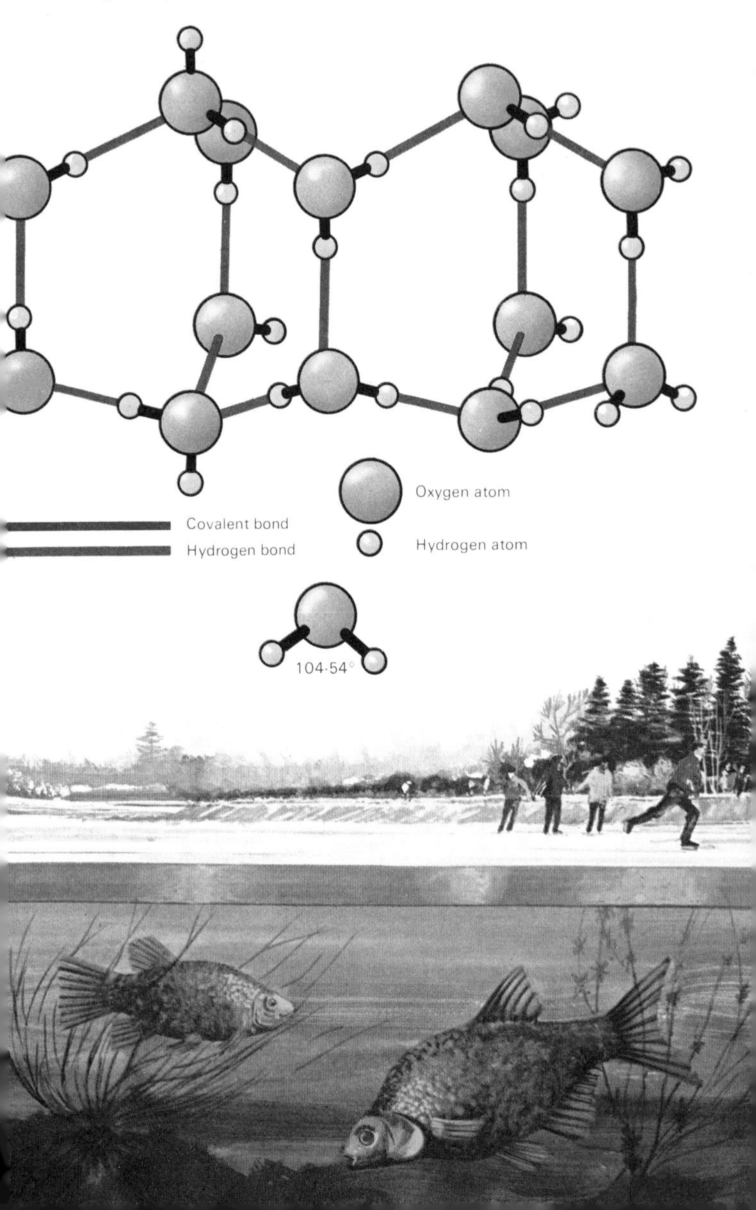

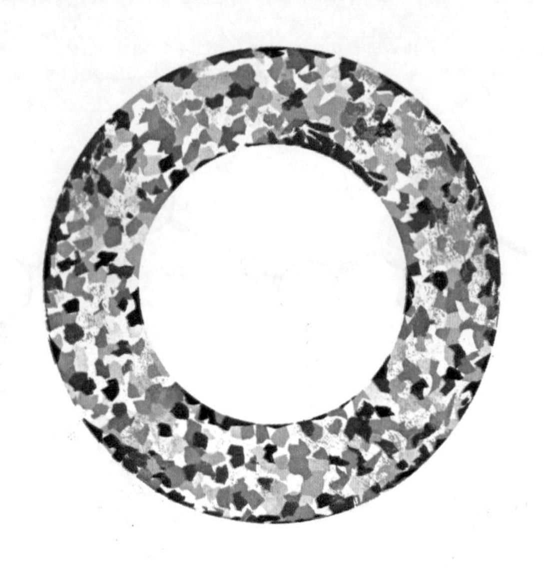

(*Above*) Lead pipe etched to show crystal grains
(*Below*) Cubic crystal lattice like that of common salt (NaCl)

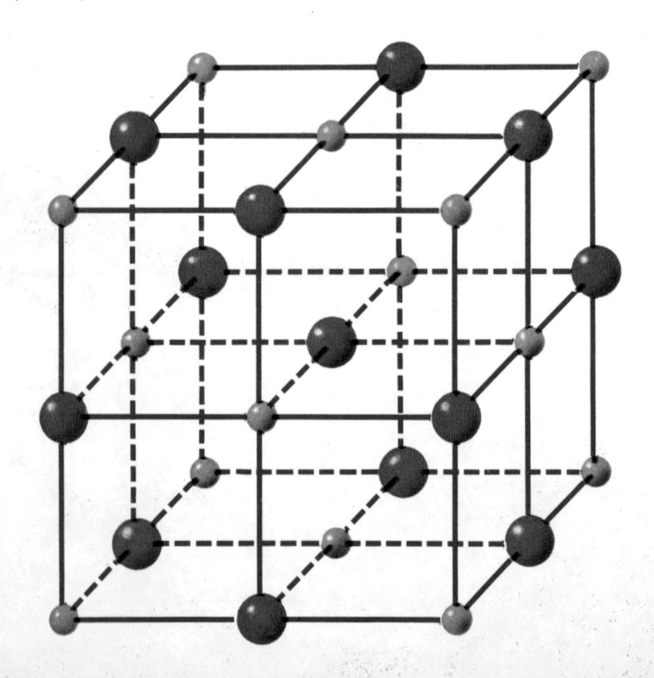

Solids

Like ice, most solids have a rigid crystalline structure. This means that they have a definite *melting point* – the temperature at which the ions or atoms making up the crystals have sufficient energy to break free from the bonds holding the structure together. There is, however, a class of solids, called *amorphous* solids, that do not have a crystalline structure. These substances, of which glass is an example, do not have a definite melting point – they become more and more plastic as they are heated, until they eventually become liquid. For many purposes it is convenient to think of an amorphous solid as a *supercooled* liquid.

Crystals in solids are of three main kinds, *electrovalent crystals, covalent crystals,* and *metallic crystals*. Electrovalent crystals are formed by ions held together by electrovalent bonds in a *lattice* arrangement. The ions are free to vibrate about fixed positions in the lattice. Electrovalent crystals dissolve readily in water, the bonds breaking down and releasing the ions into solution. Covalent crystals are of two types, those in which the molecule is the building block of the structure, and those in which the crystal itself is one giant molecule. Most organic compounds belong to the first type, each crystal consisting of many molecules held together by relatively weak intermolecular forces. Hard substances like diamond belong to the second type – the constituent atoms being held together by strong covalent forces. In metallic crystals, the metal atoms are ionized but as all the ions are positively charged there are no electrovalent forces between them. The ions still vibrate about fixed positions in the lattice as they do in electrovalent crystals, but the electrons that have left the outer orbits of the metal atoms are free to move about the crystal structure – the metal ions are immersed in a 'gas' of electrons.

When heated, all solids expand – this is because the vibrations of the ions and atoms in crystals have greater amplitude as the temperature increases. Each type of solid has a different *coefficient of thermal expansion* (the amount that unit length of a substance will expand for a rise in temperature of 1°C). This effect has to be taken into account when building bridges and laying railways.

HEAT

Until the eighteenth century there was some confusion between *heat* and *temperature*. The distinction, first correctly made by Joseph Black (1728–99), is now clearly understood in terms of the atoms and molecules of a substance. Temperature is a measure of the average energy of the atoms and molecules, whereas heat is a measure of their *total* energy. Thus temperature is independent of the quantity of matter present but heat is not. An incandescent fragment of metal from a grindstone has a high temperature but it does not burn your skin because, having only a tiny mass, it contains very little heat. An iceberg, on the other hand, has a low temperature but contains considerable heat because there is so much of it.

When a body is heated its atoms and molecules gain kinetic energy and its temperature rises. But not all kinds of matter respond in the same way; some substances increase their temperature more than others for a given quantity of added heat. Every substance has its own characteristic *heat capacity* – that is the amount of heat required to raise its tem-

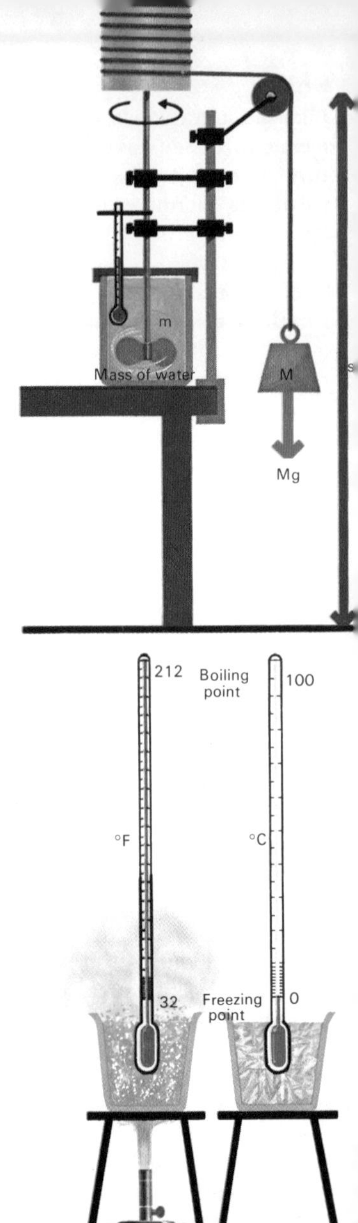

perature by 1°C. The *specific heat capacity* is the heat required to raise 1 kg of a substance by 1°C, and the quantity of heat, q, that has to be added to a body of specific heat capacity c and mass m to raise its temperature by T°C is given by $q = mcT$.

Heat can be transferred from one place to another in three ways: *conduction, convection,* and *radiation*. Conduction occurs when two solids, liquids, or gases are in contact and results from the transfer of kinetic energy between the atoms and molecules of the two substances. Convection only occurs in liquids and gases; it consists of circulating currents set up when differences in temperature cause differences in density. Radiation can occur between two masses separated by empty space.

Thermodynamics

The study of the laws involving heat changes and the conversion of heat into other energy forms is called *thermodynamics*. The *First Law of Thermodynamics* states that heat is

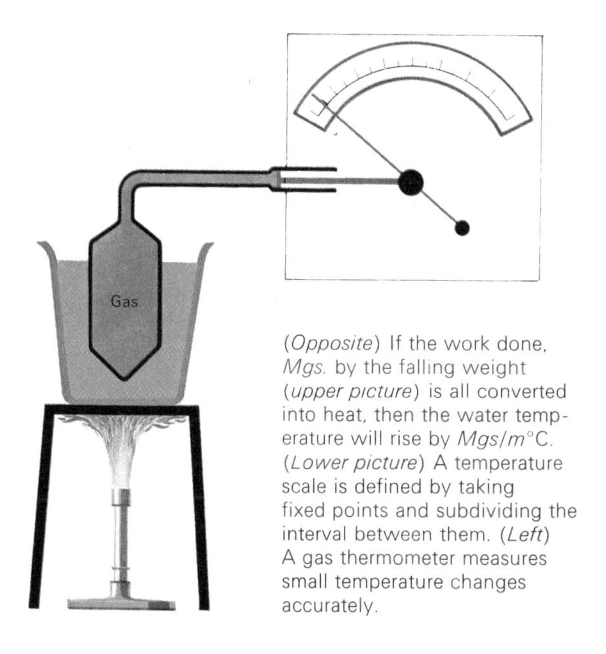

(*Opposite*) If the work done, *Mgs*. by the falling weight (*upper picture*) is all converted into heat, then the water temperature will rise by Mgs/m°C. (*Lower picture*) A temperature scale is defined by taking fixed points and subdividing the interval between them. (*Left*) A gas thermometer measures small temperature changes accurately.

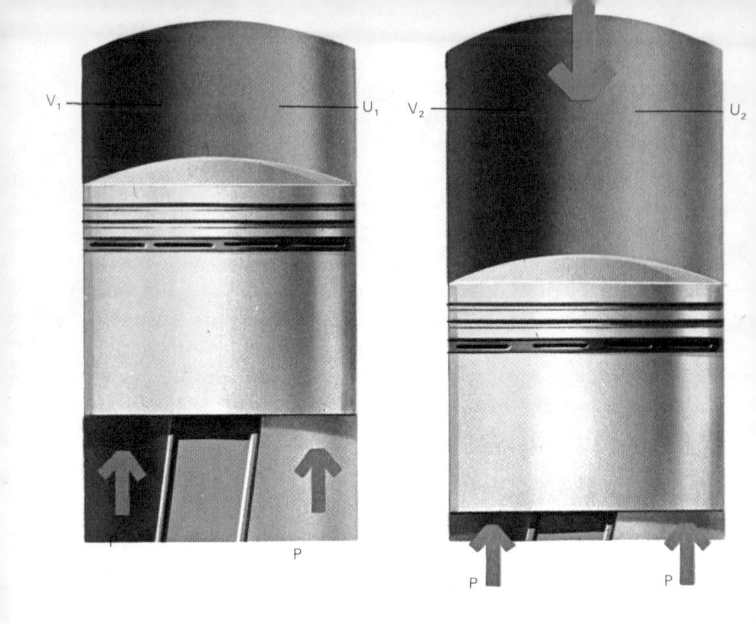

Addition of heat, q, to the gas in the combustion chamber increases its internal energy from U_1 to U_2 and expands it from V_1 to V_2 against atmospheric pressure, p.

Oxygen and acetylene burn together to release heat energy.

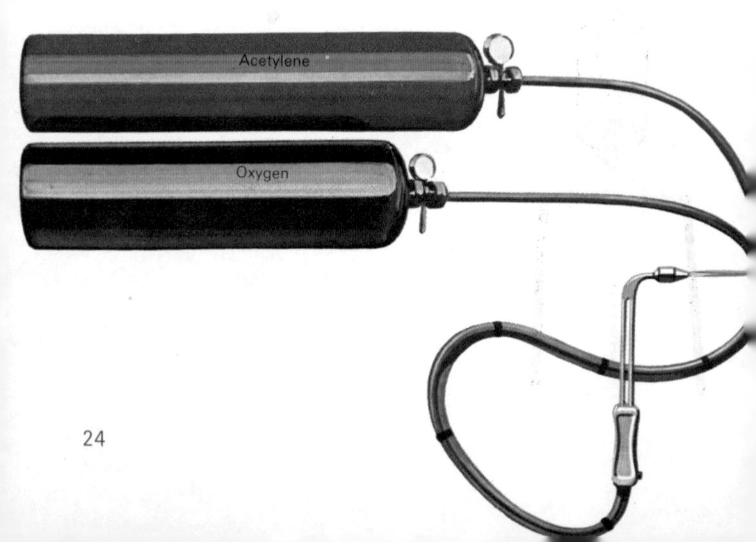

a form of energy and as such cannot disappear – it must be converted into some other form of energy. If a quantity of heat, q, is added to a system and if a quantity of work, w, is done on the system, there will be a change in the *internal* or *thermodynamic energy, U,* of the system: $U_2 - U_1 = q + w$. In a heat engine, heat is added to the system but work is done *by* the system rather than on it. In this case w is taken as negative. If the process takes place at constant heat ($q = 0$) it is said to be *adiabatic* and then $U_2 - U_1 = w$.

Usually the absolute value of U is not known, but changes in its value can be calculated by measuring q and w. If the system undergoes a change of volume $V_2 - V_1$, the work done *by* the system is $-P(V_2 - V_1)$ where P is the pressure of the environment. This quantity is positive if work is done *on* the system.

In the case of a chemical reaction, q is the heat evolved by, or added to, the reaction. If H_1 is the initial heat content of a system and H_2 its final heat content, then $q = H_2 - H_1$. H is called the *enthalpy* of the system and $H_2 - H_1 = (U_2 - U_1) + P(V_2 - V_1)$. When a chemical reaction evolves heat (an *exothermic* reaction) the enthalpy change, $(H_2 - H_1)$, is taken as negative; when heat is absorbed from the environment (an *endothermic reaction*) it is positive. In general, the convention is that heat put into a system, or work done on a system by the environment is positive, heat or work coming out of the system into the environment is negative.

When a substance changes its state, from liquid to gas, there is an *isothermal* change of energy (that is a change of energy at constant temperature) and as the change is accompanied by an expansion, work is done on the surroundings, therefore; $H_g - H_l = (U_g - U_l) + P(V_g - V_l)$. The quantity $H_g - H_l$ is called the *enthalpy of evaporation* or *vaporization* (or the specific latent heat). Thus to convert 2 kg of water at 0°C into steam at 100°C two quantities of heat are involved:

(1) The heat (mcT) required to change the temperature of the water from 0 to 100°C, namely $2 \times 100 \times c$ joules (where c is the specific heat capacity of water).

(2) The heat $m(H_g - H_l)$ required to convert the water to steam at the same temperature, namely $2 \times (H_g - H_l)$ joules. There is a similar change of enthalpy, the enthalpy of melting,

$H_l - H_s$, (the latent heat of fusion) when a solid becomes a liquid.

According to the First Law of Thermodynamics, when a system changes from one energy state, *X*, to another, *Y*, the

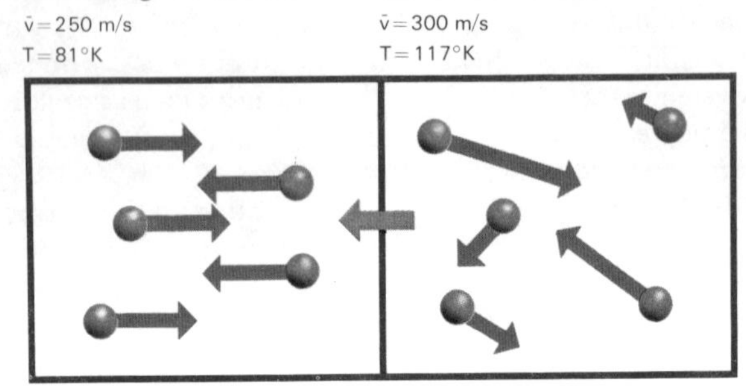

Interaction between the gases in the two compartments evens out the temperature difference between them.

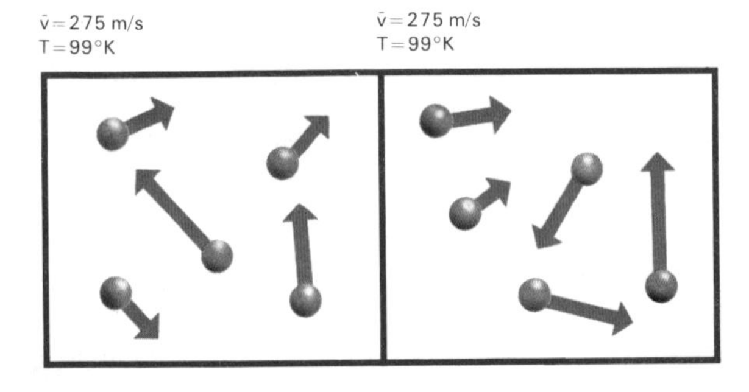

condition $U_y - U_x = q + w$ applies irrespective of the path from *X* to *Y*. However, there is one important restriction to all heat transfer processes. At no stage in the path from *X* to *Y* can heat be transferred from a source at a lower temperature to one at a higher temperature without the application of external work – heat always flows from hot to cold. This is the *Second Law of Thermodynamics*.

Imagine an insulated vessel with two compartments both containing equal volumes of oxygen. On one side, A, by some highly improbable freak of nature all the molecules have approximately the same velocity, say, 250 m/s. As temperature (in kelvins) is proportional to the average kinetic energy of the molecules (actually it is $4.8 \times 10^{22} \times$ kinetic energy) the temperature can be calculated; it is 81°K. In the other half of the vessel the molecules are moving in all directions with a wide range of velocities, the average being 300 m/s; this is equivalent to 117°K. Heat will flow from B to A, across the partition separating the two compartments, in accordance with the second law. There will, therefore, be an increase in the average velocity of the molecules in A, a consequent increase in the number of collisions between them, and therefore an increase in their *disorder* (Fig. 20b). This is another way of stating the second law: heat flows in the direction that will increase the overall disorder of a system.

The disorder of a system is measured by a quantity called *entropy, S*. The increase in disorder $S_2 - S_1$ is defined as the sum of all the increments of heat transferred at a given temperature, $S_2 - S_1 = q/T$. For example, when a liquid becomes a gas at its boiling point, there is an increase of disorder amongst the molecules – i.e. an increase in entropy $S_g - S_l = (H_g - H_l)/T$, where $H_g - H_l$ is the enthalpy of evaporation (see page 25) and T the boiling point.

This tendency for entropy to increase in all natural processes means that the energy of the universe is becoming more disordered – less available for use. Eventually there will be no available energy – a state of maximum entropy: this state is sometimes called the *heat death of the universe*.

At the opposite end of the entropy scale is the absolute zero of temperature; at this temperature all the atoms and molecules of matter would cease to vibrate or move. There would therefore be no collisions and no disorder – their entropy would be zero. This is the *Third Law of Thermodynamics*; at absolute zero the entropy of a system would be zero.

Heat engines

A heat engine is a device for converting heat energy into mechanical energy. These engines obtain their heat energy from the combustion of a fuel in which chemical energy is stored. The table shows the *energy density* of various types of fuels and energy storage devices expressed as the energy in joules stored by a fuel or device weighing 1 kg (see page 31).

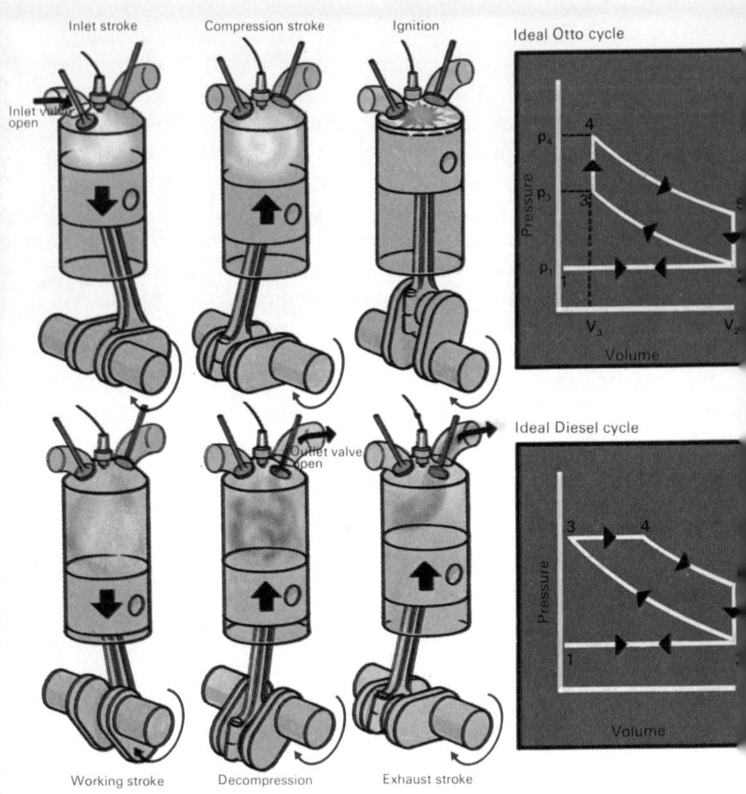

The four-stroke cycle with pressure/volume changes during Otto and Diesel cycles

The first requirement of a heat engine is that it should be as efficient as possible; the overall or thermal *efficiency* is the work output divided by the heat input (w/q). Other requirements are that it should be as cheap, light, and quiet as possible and should cause minimum pollution.

Heat engines operate on a cycle, that is a number of operations that are repeated over and over again. In these cycles a quantity of heat q_1 is absorbed at the maximum temperature T_{max}, and a smaller quantity of heat q_2 is discharged at a lower temperature T_{min}. As the work done is equal to $q_1 - q_2$ (see the first law of thermodynamics – there is no change in internal energy), the efficiency $w/q_1 = (q_1 - q_2)/q_1$

$=1-q_2/q_1$. Thus for maximum efficiency, q_2 must be as small as possible and q_1 must be as great as possible. As q_1 is proportional to T_{max}, the higher the working temperature the higher the efficiency.

The petrol engine operates on an *Otto cycle* (named after Nikolaus Otto, 1832–91). The ideal Otto cycle plotted as a graph of pressure against volume is shown with the corresponding movements of the piston and valves. This ideal cycle is not precisely followed in practice, but it illustrates the principles involved. It consists of six operations involving four piston strokes; it is therefore called a four-stroke cycle.

In operation 1–2, the piston descends, drawing in the air and petrol vapour through the open inlet valve. In 2–3, the piston rises, compressing the gas – ideally without heat loss or gain (adiabatically). During 3–4 the sparking plug sets off the explosion in the gas resulting in the addition of heat q, and the pressure in the cylinder rises sharply, from p_3 to p_4, before the piston moves. In 4–5, the working stroke, the piston is pushed down by the expanding gas: at the bottom of its stroke, 5–2, the exhaust valve opens and the pressure drops back to atmospheric pressure, P_1. The final stroke, 2–1, pushes the exhaust gas out of the cylinder.

The maximum theoretical efficiency of an Otto cycle engine is about 67 per cent, based on a *compression ratio* (the ratio of the pressure immediately before combustion to the

Comparative cost, weight and pollution of different heat engines

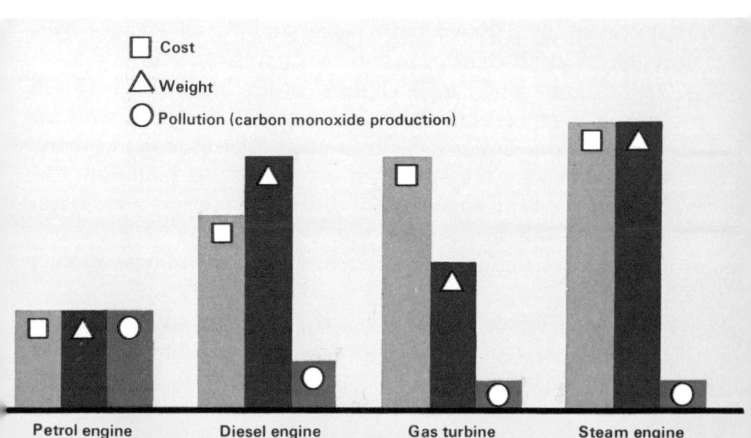

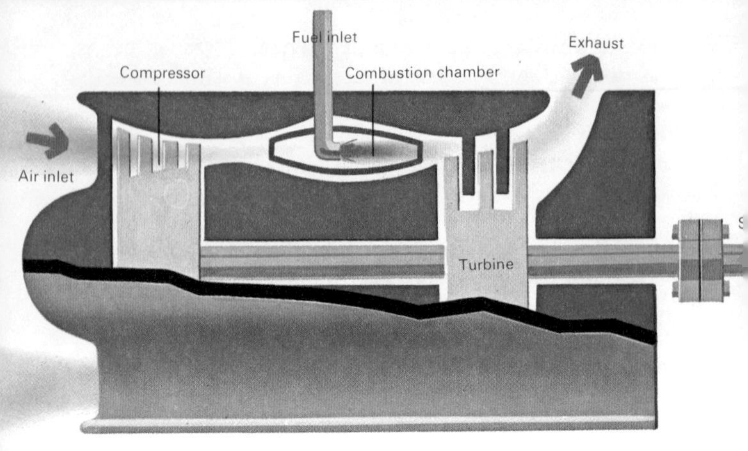

The gas turbine

atmospheric pressure $-v_2/v_3$) of 9 to 1. Higher compression ratios are not possible as the petrol vapour tends to ignite before the spark (preignition). In the Diesel engine, based on the *Diesel cycle* (named after Rudolph Diesel, 1858–1913), the air is compressed adiabatically during 2–3 until the temperature is high enough to ignite the oil being sprayed into the cylinder – no spark is required. The combustion then takes place at constant pressure, the piston moving down during the process – a horizontal line, 3–4 on the diagram, instead of the vertical line in the Otto cycle. Even though compression ratios of 14 or 15 to 1 are used with Diesel engines, their maximum theoretical efficiency is limited to 64 per cent. However, in practice both petrol and Diesel engines have much lower efficiencies than these figures.

Both Otto and Diesel engines suffer from drawbacks of limited compression ratios and limited maximum working temperatures. Because combustion is intermittent and therefore incomplete they also cause considerable pollution. The *Wankel engine,* using a rotary piston, has certain advantages of weight and smoothness, but it still has the same thermodynamic restrictions and inherent pollution problems as other piston engines.

The gas turbine breaks free from these restrictions – the compression ratio can be 30:1 and the maximum working temperature as high as 1200°C. As combustion is continuous, pollution is much less. Air is drawn into a rotary compressor and passed to a combustion chamber in which it is heated by the burning fuel: the hot gas then expands and drives a turbine which both delivers the power and drives the compressor, which is often on the same shaft.

FUEL OR ENERGY-STORING DEVICE	ENERGY STORED joules/kg
matter: mass-energy conversion	9.0×10^{16}
liquid hydrogen	1.4×10^{8}
petrol/diesel oil	4.7×10^{7}
lead-acid accumulator	1.3×10^{5}
flywheel	2.0×10^{4}
spring	2.0×10^{2}

A twin rotor Wankel engine – air is compressed in two stages.

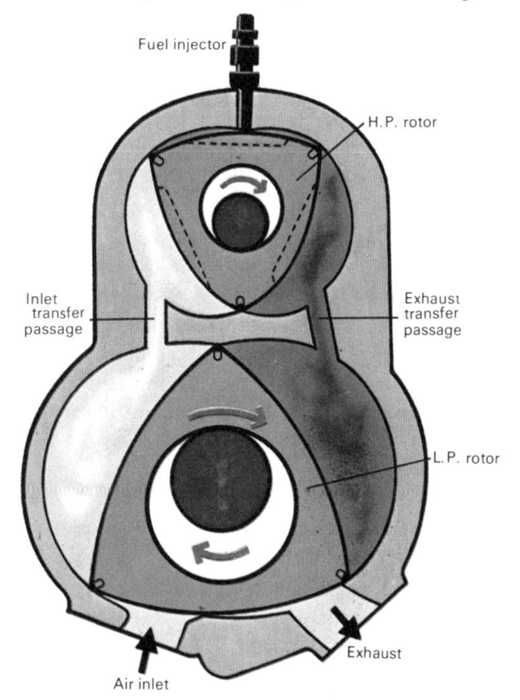

ELECTRICAL ENERGY

Chemical energy can be converted into heat by combustion and then into mechanical energy by a heat engine. Another route from chemical energy to mechanical energy is via an *electric cell* and an electric motor.

In an electric cell, chemical energy is converted directly into an electric current. This happens when neutral atoms and molecules, in which the chemical energy is stored, break down into charged fragments in solution; when the electrons liberated from the atoms flow through an external metal conductor they constitute an electric current. The illustration shows a simple type of cell (called a *Daniell Cell*) in which a zinc (symbol Zn) rod dips into a solution of dilute sulphuric acid (H_2SO_4) and a copper rod dips into a solution of copper sulphate ($CuSO_4$); the zinc rod and the sulphuric acid are contained in a porous pot which stands in the copper sulphate solution.

When the zinc dissolves in the sulphuric acid, each zinc atom breaks down into two electrons and a positively charged zinc fragment called an *ion* (an atom that has lost or gained one or more electrons and is no longer neutral – in this case the atom loses the two electrons in its outer shell). The reaction is written: $Zn \rightarrow Zn^{2+} + 2e$. The sulphuric acid also breaks down into ions in solution: $H_2SO_4 \rightarrow 2H^+ + SO_4^{2-}$. In the same way the copper sulphate breaks down: $CuSO_4 \rightarrow Cu_2^+ + SO_4^{2-}$. The two electrons liberated on the zinc rod flow through an external wire to the copper rod. Here they combine with copper ions from the copper sulphate solution $-Cu_2^+ + 2e \rightarrow Cu-$ to deposit more copper on the copper rod. The overall reaction of the cell is therefore: $Cu_2^+ + Zn \rightarrow Cu + Zn^{2+} + \textit{energy}$. As the copper atoms and zinc ions have less internal energy than the copper ions and zinc atoms, energy is given out; in this case the work done by this energy is to make electrons flow through an external metal wire.

The rods in a cell are called the *electrodes* and the solution in which they stand is called an *electrolyte*. The negative ions in the electrolyte (the sulphate ions SO_4^{2-}) are attracted to the positive electrode called the *anode,* and the positive ions (Zn^{2+}, Cu^{2+}, H^+) are attracted to the negative electrode called

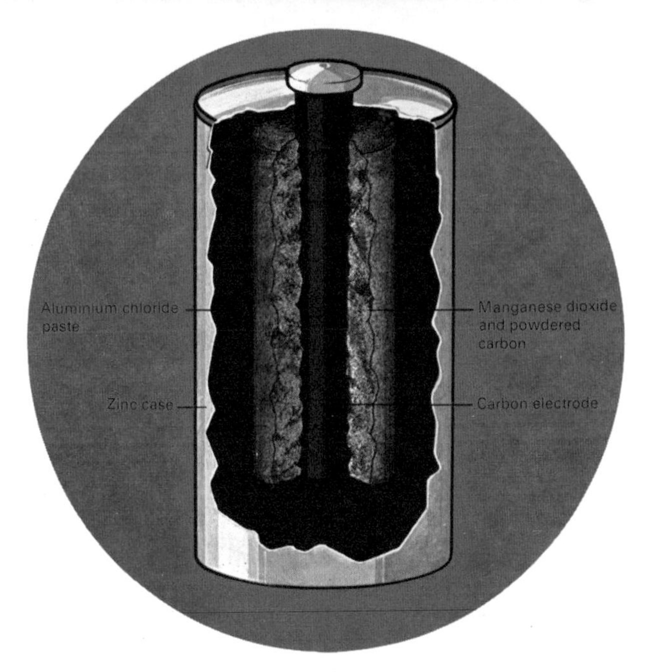

(*Above*) Dry battery. (*Below*) Daniell cell

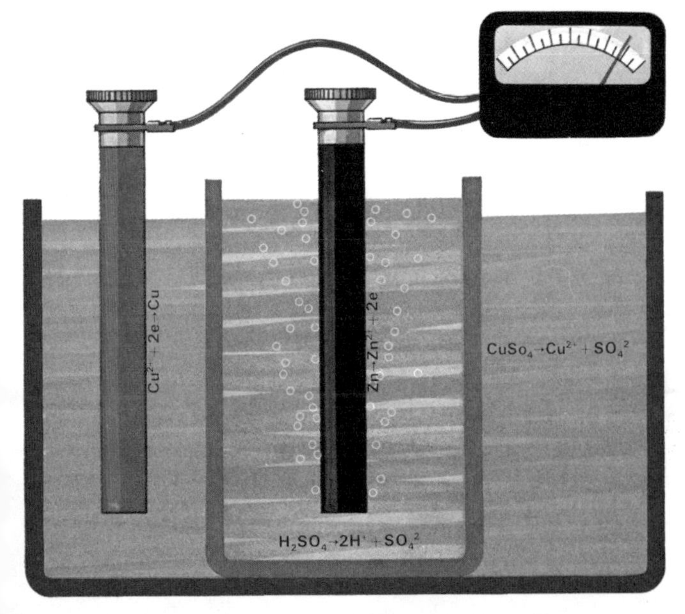

Chromium plating by electrolysis

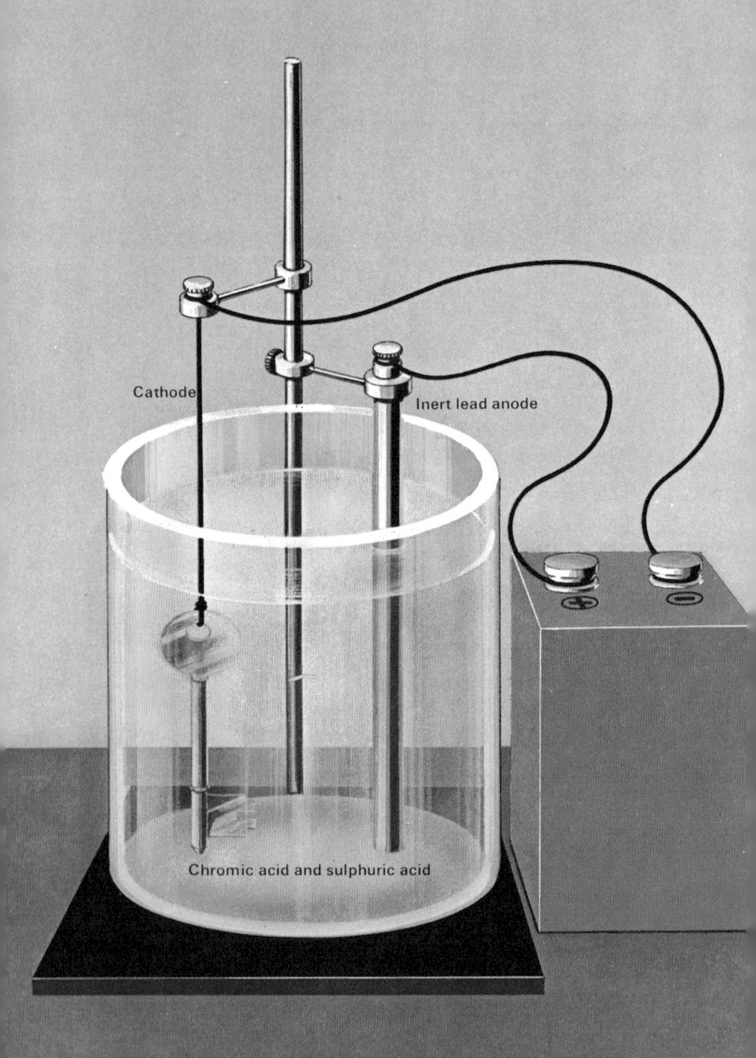

the *cathode*. The current of electrons flowing through the external wire is equal to the sum of the negative and positive ion currents flowing through the electrolyte. The unit of current is called the *ampere* and 1 ampere is equal to a flow of about 6×10^{18} electrons per second. The unit of electric charge is called the *coulomb*, and 1 ampere is equivalent to a flow of 1 coulomb of charge per second. Therefore the coulomb is equivalent to the charge of 6×10^{18} electrons and each electron has a charge of $\frac{1}{6} \times 10^{-18} = 1.6 \times 10^{-19}$ coulombs. In the SI system the basic electrical unit is the ampere, the coulomb being defined in terms of the ampere.

The ratio of the energy produced by a cell to the quantity of charge separated (in joules per coulomb) is called the *electromotive force* (emf) of the cell and it is measured in *volts*, 1 volt being equal to 1 joule per coulomb. For the Daniell cell a voltmeter placed across the electrodes would show 1.1 volts – this is called the *potential difference* (p.d.) of the cell.

When all the copper ions from the copper sulphate solution have been used up (or all the zinc of the zinc electrode dissolved, current will cease to flow. However, the whole process can be reversed by placing another source of emf (such as another cell) across the terminals. This makes all the chemical equations go into reverse, the copper electrode dissolving and zinc being deposited. This is called *electrolysis*.

Electrolysis is also used to deposit one metal on another. If an object, made of a relatively cheap metal, is made an electrode in an electrolytic cell, it can be plated with a more expensive metal such as chromium, silver, or gold. This is called *electroplating*.

The Daniell Cell is only one of many different types; the common dry battery used in torches and radios is usually a form of *Leclanché cell*. This type of cell is not suitable for recharging.

Ohm's Law

The flow of current through a wire is similar to the flow of water through a pipe – the quantity of water flowing per second being analogous to the number of electrons flowing per second, or the current. The pressure head required to keep the flow going on this analogy is similar to the emf or voltage. Just as there is friction between the walls of the pipe and water flowing through it, so there is a *resistance* to the flow of electricity through a conductor. This is due to the electrons colliding with the atoms of the conductor thereby losing some of their energy. The unit of resistance, called the *ohm,* is chosen so that 1 ohm is the resistance to the flow of a current of 1 ampere when the potential difference is 1 volt. In general, the potential difference, V, divided by the current, I, is equal to the resistance, R, or $V=IR$. This is *Ohm's Law* and it enables the characteristics of electric circuits to be calculated.

An electric light bulb consists of a short length of thin, high-resistance wire (the filament) in a glass bulb containing an inert gas such as argon. From the mains voltage and the current taken, the bulb's resistance and power consumption can be found (see illustration).

When a current of electrons flows through a wire the continual collisions of the electrons with the atoms of the wire cause these atoms to vibrate more energetically, their average energy is therefore increased and the temperature of the conductor rises. This is why the filament in the bulb glows with white heat. The energy, q, (in joules) dissipated by an electric current is equal to the voltage multiplied by the current multiplied by the time, t, for which it flows, that is, $q=VIt$, or using Ohm's Law $q=I^2Rt$.

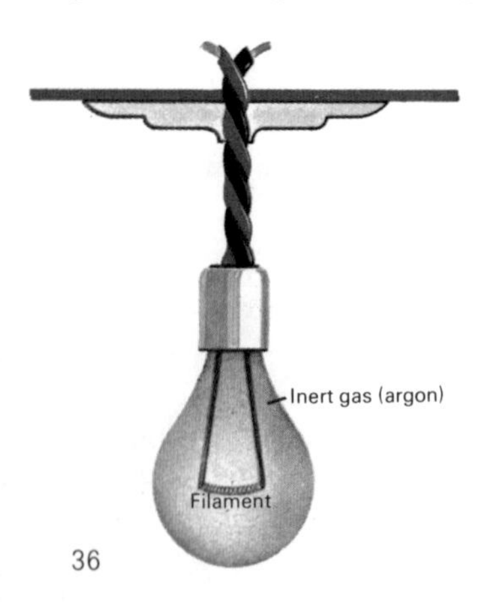

(*Left*) The light bulb runs from a 220 volt supply, taking a current of 0.45 amps. Its resistance is therefore 220/0.45=490 ohms and its power consumption is 220×0.45=100 watts.

(*Opposite*) Flow of current driven by a voltage difference is analogous to flow of water under a pressure head.

The *power* of an energy source is the rate at which it is capable of doing work. The unit of power is the *watt,* which is equal to 1 joule per second. The power in watts of an electric device is equal to the voltage times the current flowing through it, which is VI or I^2R. The unit for charging consumers for electricity is the kilowatt-hour, 1000×3600

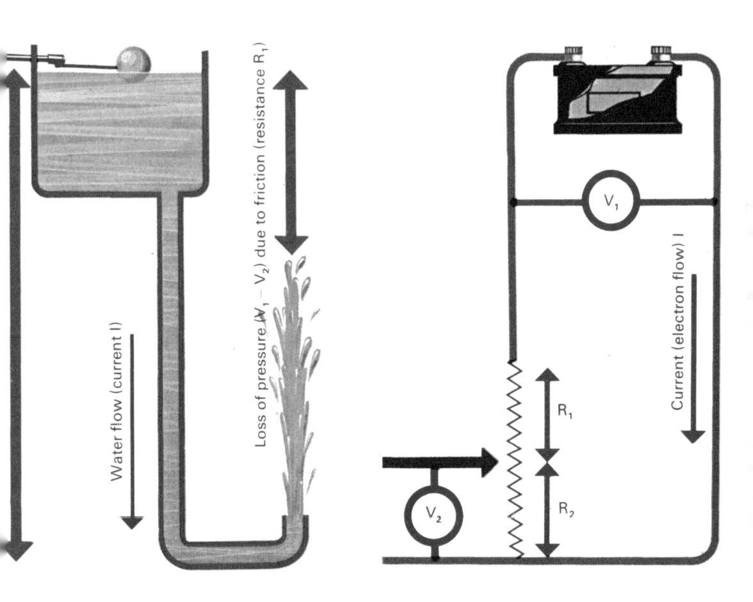

watt–seconds; as 1 watt–second is equal to 1 joule, 1 kilowatt–hour $= 3.6 \times 10^6$ joules.

The resistance of a conductor increases with temperature; the more energetic the vibrations of the atoms, the more collisions there are with the electrons flowing through the conductor. Conversely the resistance decreases with tempera ture, but at a very low temperature, within a few degrees of absolute zero, resistance disappears altogether. The atomic vibrations are so feeble that very few collisions with electrons occur: the conductor is then called a *superconductor.*

Capacitance

Electric charge can be stored in an electric circuit by a device called a *capacitor*. The simplest form of capacitor consists of two parallel metal plates separated by an insulating medium (called a *dielectric*), such as air or mica. If the two plates are connected to a battery, one plate (connected to the positive terminal) acquires a positive charge and the other plate a negative charge; there is therefore a potential difference, V, between the plates. The ratio of charge stored, Q, to this potential difference is a quantity known as *capacitance*, C, which depends only on the characteristics of the capacitor; for a parallel plate capacitor, $C = A\varepsilon/d$, where A is the plate area, d is the distance between the plates, and ε is a constant associated with the dielectric, called the *permittivity*. Capacitance is measured in *farads*: if the potential difference across the plates is 1 volt and 1 coulomb of charge is stored, the capacitance is 1 farad; this is a very large quantity and the microfarad is used in practice.

Electricity is concerned with the movement of electric charge; *electrostatics* studies the properties of electric charges at rest. Electric charge is due to the presence (negative charge) or absence (positive charge) of electrons. If electrons are removed from the atoms of a material – by friction, for example – it acquires a positive charge. If glass is rubbed with

Series and parallel addition of resistance and capacitance

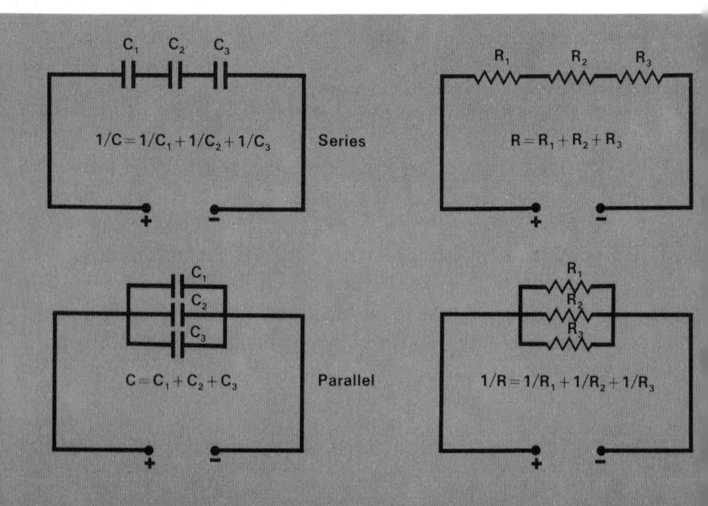

a silk cloth, electrons are detached from the glass atoms and are transferred to the silk; the glass becomes positively charged, the silk, negatively charged. A positively charged body will attract a negatively charged one; two positive or two negative surfaces will repel each other. The force of attraction or repulsion is related, by *Coulomb's Law,* to the product of the charges, $q_1 q_2$, divided by the square of the intervening distance.

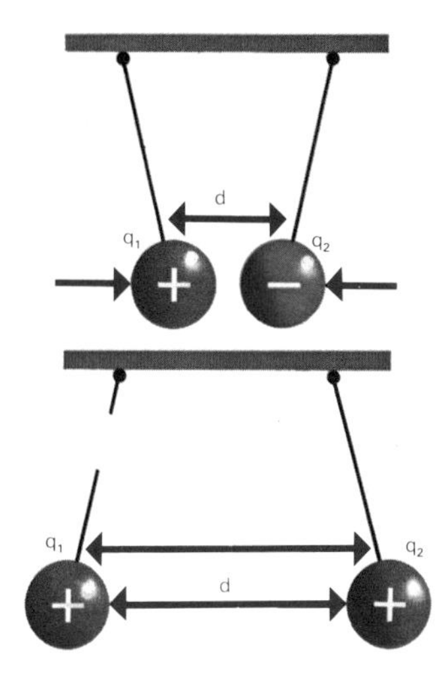

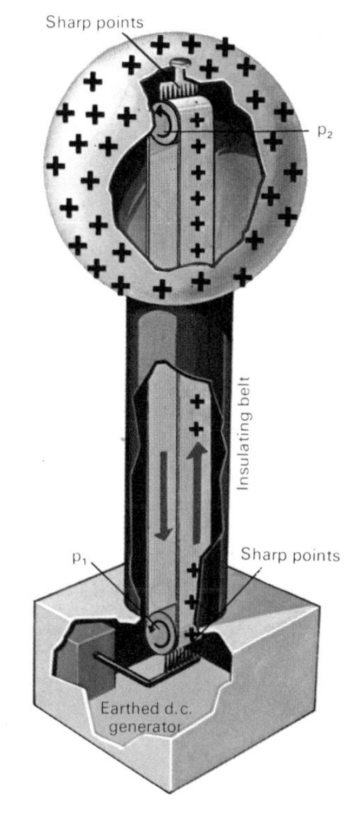

(*Upper*) Opposite charges attract, like charges repel. (*Right*) The insulated metal sphere of a *Van de Graaf* generator accumulates electric charge conveyed by the moving belt. Ionization of air molecules at the lower set of points sprays positive charge onto the belt. The upper points convey the charge to the sphere.

Magnetism

Forces of attraction and repulsion exist not only between electrically charged bodies but also between magnetic substances such as lodestone, a naturally occurring ore of iron oxide. The magnetism appears to be concentrated at two small areas near the ends of the magnet; these are the *magnetic poles*. When a bar magnet is freely suspended, it settles so that the same end tends to point in a northerly direction. The pole at this end is therefore called the *north pole*; the pole at the opposite end of the magnet is the *south pole*. A north pole will attract a south pole; like poles repel each other. A magnetic pole cannot exist in isolation as an electric charge can, but only in conjunction with another pole, as a *dipole*.

The region surrounding the magnet in which the magnetic force has an appreciable value is regarded as the *magnetic field* of the force. An electric field denotes the region in which an electric force acts. When two fields overlap there is said to be an interaction. With a magnetic field, definite *lines of force* exist that define the direction of the field; these lines

Lines of force around two magnets. The force of attraction or repulsion between two poles, M_1 and M_2, a distance, d, apart is $M_1M_2/\mu d^2$ where μ is the permeability of the medium.

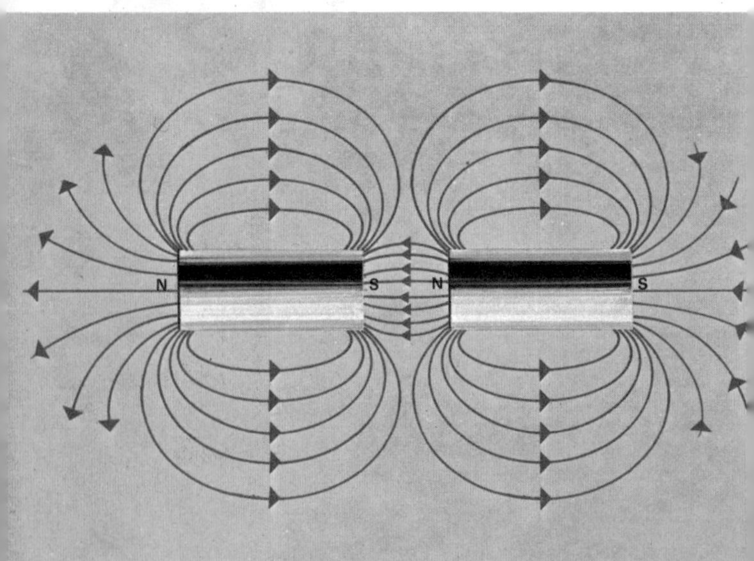

originate at the north pole and converge at the south pole.

Some metals, such as iron and nickel, and certain alloys are considerably more magnetic than other substances, and are said to exhibit *ferromagnetism*. A ferromagnet does not usually occur naturally in a magnetic state but is made a

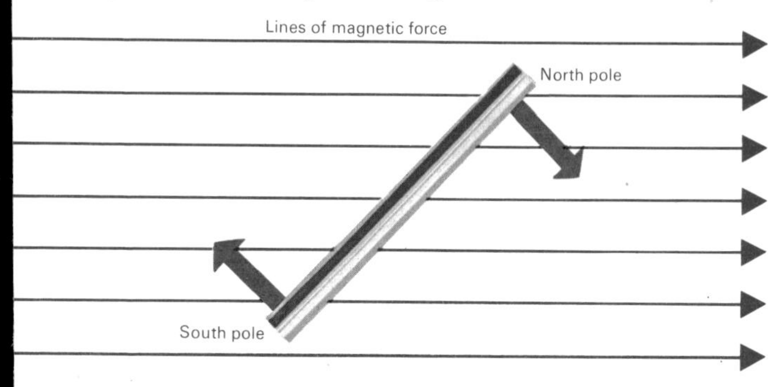

Each pole of a bar magnet experiences an equal but opposite turning force when placed in a magnetic field.

magnet by a variety of methods, which include placing it in the magnetic field of an electromagnet. When the magnetic field is removed, some ferromagnets retain their magnetism to a marked degree (*hard* magnets) others lose it fairly rapidly (*soft* magnets). Above a certain temperature, called the *Curie temperature,* whose value depends on the material, a ferromagnet loses its magnetic powers and becomes *paramagnetic* with only very weak magnetic properties.

Magnetic domains

A magnet can be broken down into smaller and smaller magnets. This fact, together with the knowledge that most ferromagnets must be magnetized and can also be demagnetized – by heating for example – led to the idea that a ferromagnet contains an enormous number of tiny elementary magnets. These component magnets are large groups of atoms coupled together very strongly by the complicated interactions produced by the motion of their electrons – not only orbital motion but also a rotation about the electron axis, known as *spin*. This coupling makes each group behave as a tiny magnet; these groups are called *domains*. Each domain contains about 10^{15} atoms, 6000 domains occupying the size of a pinhead. Before a ferromagnetic crystal becomes magnetized, the magnetic field of each domain can point in any of several directions such that the net magnetism of the substance is zero. When an external field is applied, the elementary magnets tend to become aligned in the direction of the field and the substance is then magnetic.

The earth's magnetic field

A freely suspended magnet always comes to rest with its north pole pointing in a northerly direction. A magnet has been used for centuries, in the form of a *compass*, to find and maintain direction in navigation. As a north pole is attracted to a south pole, there must be a magnetic field associated with the earth, which in fact behaves as if there were a huge bar magnet at its core, aligned approximately north-south. The origin of the earth's magnetic field is at present unknown. The temperature of the iron-nickel core of the earth is far above any Curie temperature so that ferromagnetism in the core is not the cause. Magnetism can also be produced by an electric current and it is thought that small electric currents are generated by movement of the earth's fluid core.

Magnetic effect of an electric current

The fields of electricity and magnetism are intimately connected. In 1819, Hans Oersted showed that the flow of electric current in a straight wire produces a magnetic field concentric with the wire; the direction of the magnetic field depends on the direction of the current. If a current flows through two wires a force between the wires is produced that is attractive if the two currents flow in the same direction and repulsive if the flow is in opposite directions. The *ampere* is now defined as the current, flowing in two infinitely long, straight, parallel wires placed 1 metre apart in a vacuum, that

(*Upper*) Effect of applied magnetic field on domains in a ferromagnetic crystal. The randomly oriented domains (*a*) become lined up with the external field (*b* and *c*).
(*Lower*) Auroral displays at a height of about 300 km are manifestations of the Earth's magnetic field.

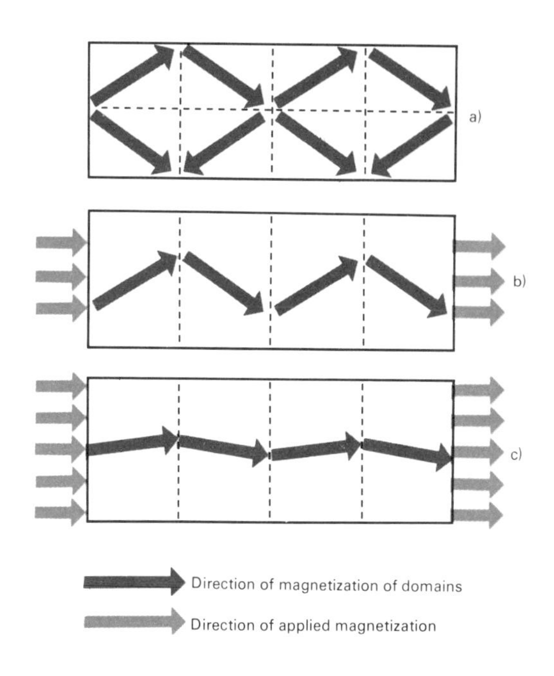
a)
b)
c)
Direction of magnetization of domains
Direction of applied magnetization

would produce a force of 2×10^{-7} newtons per metre of length.

In a circular coil of wire the magnetic lines of force are more concentrated inside than outside the coil, the field being similar to that of a permanent bar magnet. When such a coil is wrapped around a soft ferromagnetic core, it is known as an *electromagnet*. Its magnetic field is proportional to the current and the number of turns of wire, and inversely proportional to the length of the coil. While the current is flowing the ferromagnetic core becomes magnetized by the coil's magnetic field and consequently enhances it. When the current is switched off, the core loses its magnetism.

Whenever the lines of force around a current-carrying coil are changing, an electromotive force is *induced* in that coil; the magnitude of the emf is proportional to the rate of change of the magnetic field per unit area, (*magnetic flux*). This is *Faraday's Law of Induction*. The direction of the electromotive force is such that it always opposes the change producing it – *Lenz's Law*. This effect is particularly important with alternating currents (currents in which the direction of electron flow is periodically reversed – see page 48). Thus, when an alternating current is passed through a coil, the changing magnetic field induces an electromotive force in the coil, which opposes the flow of current: the induced voltage is proportional to the rate of change of current. The constant of proportionality is called the *coefficient of self inductance, L,* which is measured in *henries*: the inductance is 1 henry when a rate of change of current of 1 amp/second induces a voltage of 1 volt. Inductance coils are important elements in electronic circuits.

If an alternating current is passed through a coil (the primary) wound round a soft iron core, a voltage can be induced in a separate coil (the secondary) with which it is magnetically linked. The induced voltage causes a current to flow in the secondary which is of the same frequency, but in

The magnetic effect of an electric current. Parallel wires with currents in opposite directions repel one another (a). If the currents are in the same direction the wires attract one another (b). A current around a coil (c) creates a field like that of a bar magnet. This effect is intensified by winding a number of coils (d).

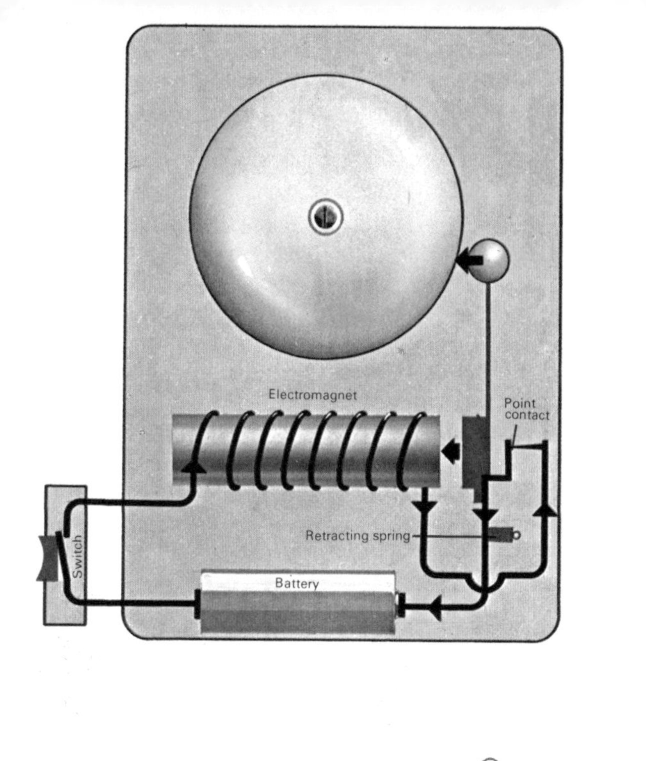
Electromagnet
Point contact
Retracting spring
Switch
Battery

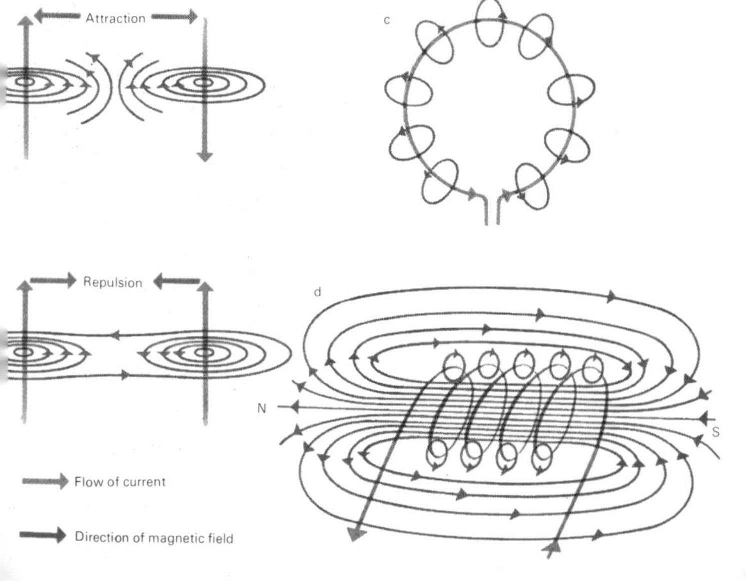
Attraction
c
Repulsion
d
N
S
Flow of current
Direction of magnetic field

the opposite direction to that in the primary. The magnitude of the induced voltage is proportional to the rate of change of current in the primary; the constant of proportionality in this case is called the *coefficient of mutual induction, M,* which like self inductance is measured in henries. This arrangement is called a *transformer* and it is used for either increasing or decreasing voltage. If an alternating current, i_1, in the primary of n_1 turns of wire induces a voltage V_2, in the secondary of n_2 turns, then, in an ideal transformer, $n_1/n_2 = V_1/V_2 = i_2/i_1$, where V_1 is the voltage in the primary and i_2 the induced current in the secondary. A *step-up* transformer increases the

Voltage transformer

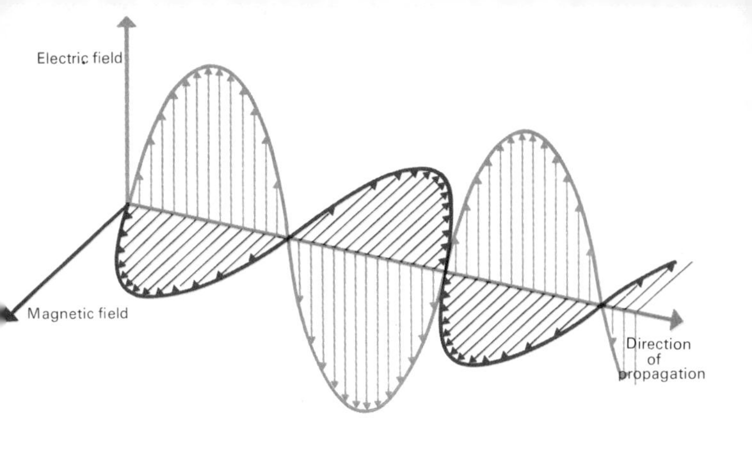

An electromagnetic wave consists of varying magnetic and electric fields at right angles to one another.

ratio V_2/V_1; a *step-down* transformer decreases the ratio. These two types of transformer play an important role in the transmission of electricity (see page 50).

The motion of any charged particle, including an electron, will produce a magnetic field. Conversely, when a charged particle enters a magnetic field, travelling at right angles to the direction of the field, it experiences a force that acts perpendicular to both particle and field direction; this force makes the particle move in a circular path, the radius of which depends on the particle velocity and the magnetic field strength. If a particle enters a magnetic field in a direction parallel to that of the field, it moves in a helical path around one of the lines of force.

In 1864, James Clerk Maxwell worked out four equations, known as *Maxwell's equations*, from which the inter-relationships between electricity and magnetism could be deduced under any conditions. These equations show that an electric field is always associated with a magnetic field. A field, resulting from the acceleration of a charged particle, is emitted in all directions: it consists of a varying magnetic field at right angles to an electric field, both of which are at right angles to the direction of motion. This is known as *electromagnetic radiation*, of which light and radio waves are examples.

Dynamos and generators

When a conductor is moved across a magnetic field a current is generated in it. Current produced in this way is the source of nearly all the electricity used domestically and industrially. A *dynamo*, or *generator*, is a device in which a continuous conductor is rotated in a magnetic field. In a simple generator a permanent magnet supplies the magnetic field and a single coil of wire is rotated in the field, the current induced in the coil being picked up by two carbon brushes pressing onto two slip rings. The maximum current is generated in the sides of the coil when it cuts the maximum number of lines of force, that is, when it is horizontal, (positions 2 and 4). When it is vertical, (positions 1 and 3) no lines of force are cut and the current is nil. As the coil rotates, each side of the coil cuts the lines of force in opposite directions and the direction of the current after half a revolution is therefore reversed. The direction of the current depends on the direction of the field

Generation of alternating current

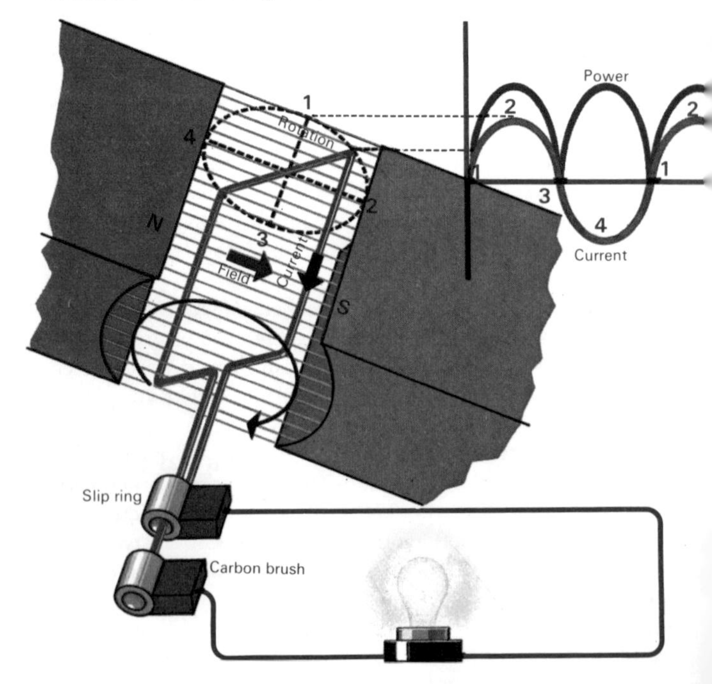

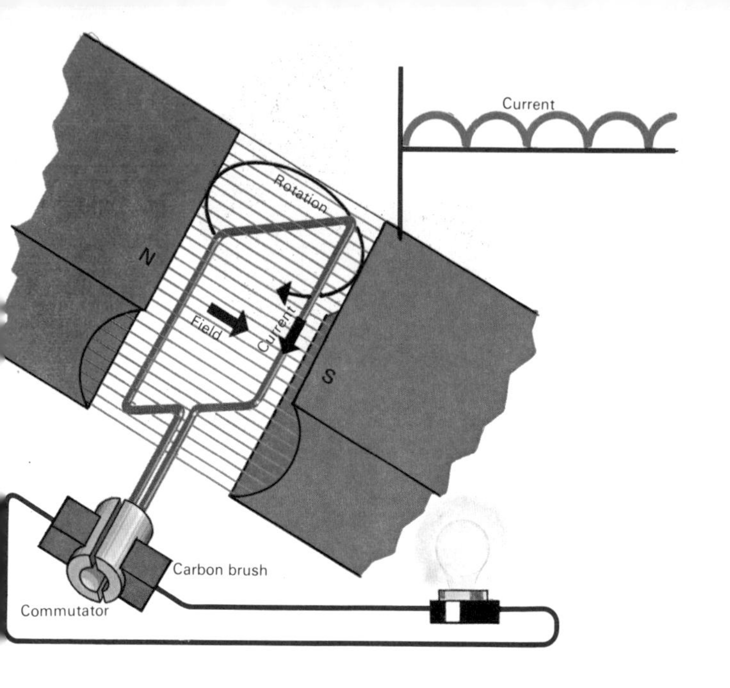

Generation of direct current using commutator

and the direction of motion. The net effect of one revolution is shown opposite.

This form of output is called *alternating current* (a.c.) and because the equation for the current is $I = I_m \sin\theta$ (where I is the current when the coil makes an angle θ with the horizontal and I_m is the maximum current), it is known as a *sine wave*. The maximum current, I_m, depends on the strength of the field and the rate of rotation of the coil.

The power of an electric current depends on the square of the current (I^2R) and because the square of a negative quantity is positive, the negative swing of the sine wave produces a positive power (the blue line on the graph). In calculating the current output of a generator it would obviously be wrong to use I_{max} as for most of the cycle the current is much less than this. The effective current, I_{eff}, works out as $I_m/\sqrt{2}$ or a little less than three quarters of the maximum.

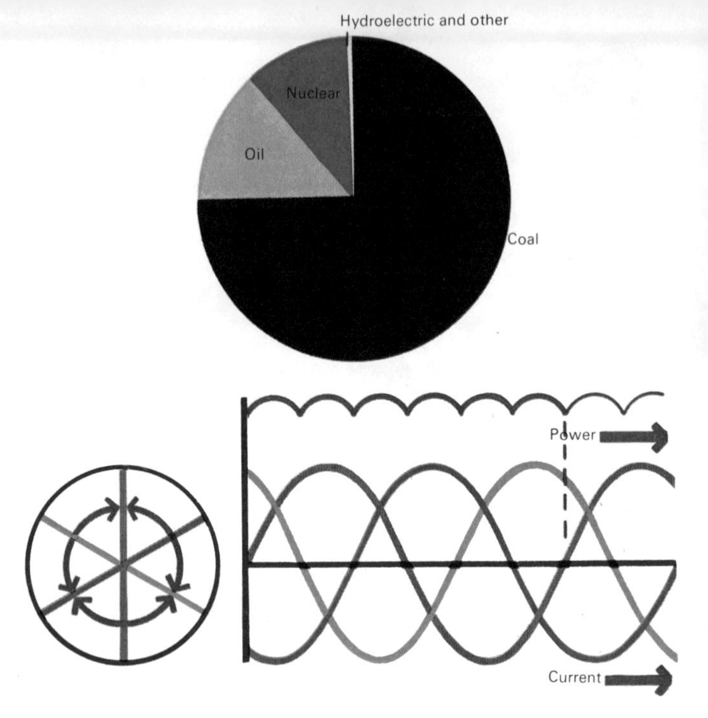

(*Upper*) Use of fuels in electricity generation. (*Left*) Angles between windings of three phase generator. (*Right*) Current and power in three phase system

If a *direct current* (d.c.) output is required from a generator, the slip ring is replaced by a split ring called a *commutator*. This device reverses the negative swing of the sine wave, thus producing a nearly smooth output. The smoothness can be improved by adding more coils and dividing the commutator into more sections.

The generation and distribution of electricity

Electricity cannot be easily stored – batteries and accumulators are essentially small-scale devices – but it can be transmitted relatively cheaply from one place to another. For this reason during the course of this century it has become the main source of energy for maintaining our civilization. Electricity for domestic and industrial use is produced by generators based on the principles described on pages 48 and 49. Over 99 per cent of these generators are made to rotate by coupling them to steam turbines, the remaining generators being rotated by water turbines (in hydroelectric schemes), Diesel engines, or gas turbines.

Because the most efficient generators produce alternating current and because alternating current can easily be changed by means of a transformer, nearly all electric power is distributed in this form. The most economical way to transmit alternating current through wires over long distances is to use a high voltage, low current, three-phase system. High voltage and low current is used because the heating loss is less (heat loss depends on the square of the current, I^2R) and because wires can be thinner. The reason for using three-phase systems is that generators with three separate windings are more efficient than those with a single winding.

Methods of generating electricity

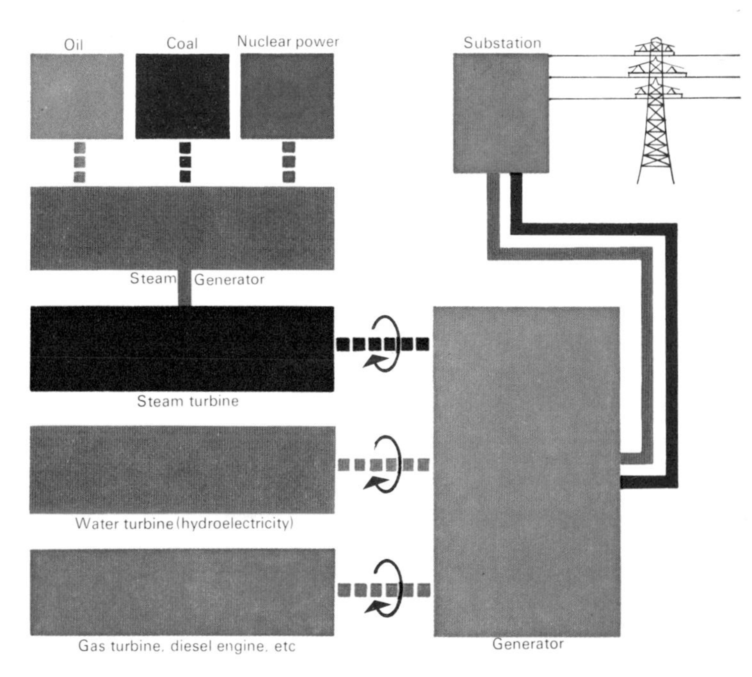

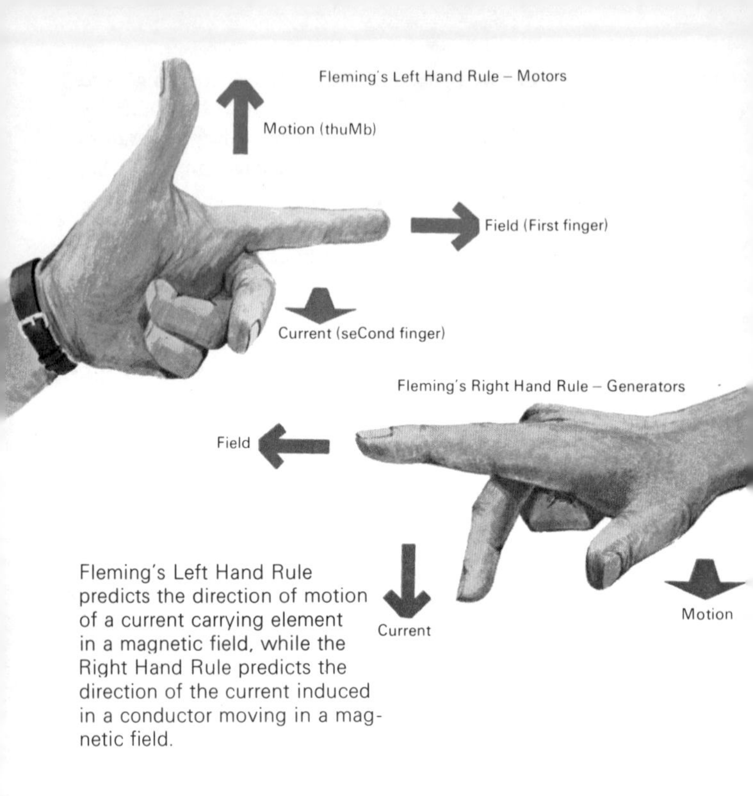

Fleming's Left Hand Rule predicts the direction of motion of a current carrying element in a magnetic field, while the Right Hand Rule predicts the direction of the current induced in a conductor moving in a magnetic field.

Electric motors

Electric motors convert electrical energy directly into mechanical energy. Some small motors in domestic devices, such as electric shavers, provide only a fraction of a kilowatt while some large industrial motors can provide a power of many thousands of kilowatts. Electric motors depend on the opposite principle to the generator: when a current flows through a conductor suspended in a magnetic field the conductor experiences a force in a direction that is perpendicular both to the direction of the current flow and the direction of the field. The illustration shows these directions for a generator, using the thumb and first two fingers of the right hand *Fleming's Right-hand Rule*), and for a motor (*Fleming's Left-hand Rule*).

Some motors are designed to run on a d.c. supply, others

on a.c., and some are *universal* (they will run on both). Motors running on d.c. are the most practical for uses in which a variation of speed is important, such as electric trains and cars. They consist of a stationary electromagnet (the *stator*) providing a magnetic field, and a number of coils wound on an armature (the *rotor*) into which current is fed through a commutator. The commutator switches the current from one coil to the next as the armature rotates.

The two main types of a.c. motor are the *synchronous motor* (in which speed is controlled by the frequency of the a.c. supply) and the *induction motor*. In the synchronous motor, the polarity of the magnetic poles in the stator alternates with the direction of the current, thus creating a rotating magnetic field which produces a force on the rotor which makes it revolve. However, as the stator field moves too fast to overcome the inertia of the rotor from rest, these machines are not self-starting – a d.c. starter has to be used. As the speed is 'controlled by the frequency of the supply, which is difficult to vary, they are used for large loads that require a constant speed, such as pumps or compressors.

Wiring of d.c. motors. The control resistance is used to vary the speed.

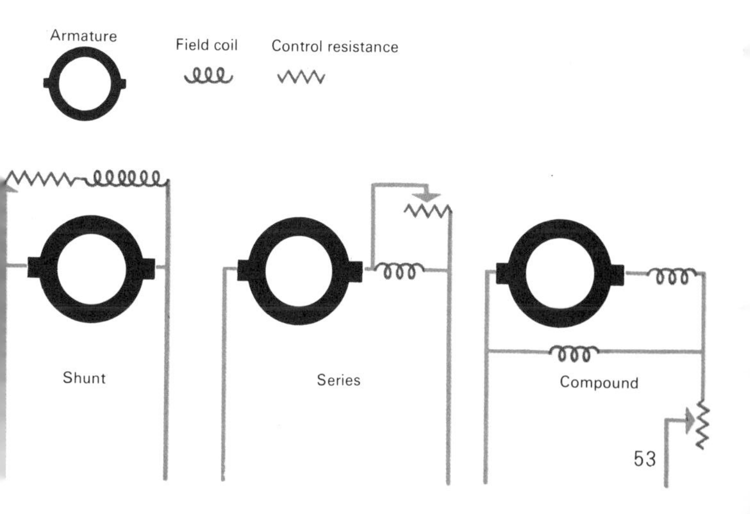

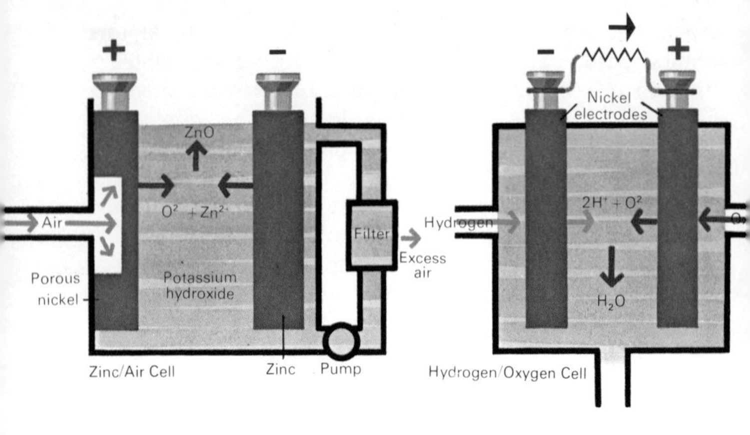

The zinc/air fuel cell (*left*) and the hydrogen/oxygen fuel cell (*right*)

Electric cars

Every year 6 000 000 tons of carbon monoxide are pumped into the air over the British Isles alone by the internal combustion engines of cars. This pollution could be reduced to zero if electric motors were to replace internal combustion engines. This has not yet happened because the problems of energy storage have not been solved. As the Table on page 31 shows, 1 kg of petrol stores nearly 400 times more energy than 1 kg of lead accumulator. This means that the energy stored in 5 gallons (about 18 kg) of petrol is equivalent to the energy stored in some 7 tons (7200 kg) of batteries. Admittedly, electric motors are more than twice as efficient as petrol engines, but even the 2 or 3 tons of batteries that would be required for the same mileage, would make a very heavy car indeed.

Two lines of research are being pursued to overcome this problem – improving accumulators and making workable *fuel cells*. The difference between an accumulator and a fuel cell is that while an accumulator stores energy, a fuel cell generates it. Improved versions of the lead accumulator have already been produced and these are able to power a 'commuter' car with a 30 mile range and a top speed of about 30 m.p.h. Another type of battery, the zinc/air battery is one of the most promising developments. This type of battery is already being used to power an experimental car – it delivers some 5 – 7 times more energy than the lead battery. Other more sophisticated accumulators being developed include the sodium/sulphur and lithium/chlorine cells; they store 10 and 15 times more energy per kilogram than the lead battery respectively, but they require a working temperature of about 300°C which is a great disadvantage.

In the accumulator, electrical energy is fed into the cells during charging, stored as chemical energy, and released as electrical energy again on discharge. In the fuel cell, fuel and oxygen are fed into the cell and their chemical energy is converted directly into electrical energy. The accumulator feeds on electricity – its chemicals do not need replenishing; the fuel cell, on the other hand, feeds on chemicals and needs no charging.

Fuel cells are extremely efficient converters of chemical to electrical energy (about 80 per cent efficient) and although they have better specific energy than lead batteries they are very bulky and occupy about the same volume. They are in use in spacecraft and it could be that they will revolutionize land transport.

The lead/acid accumulator

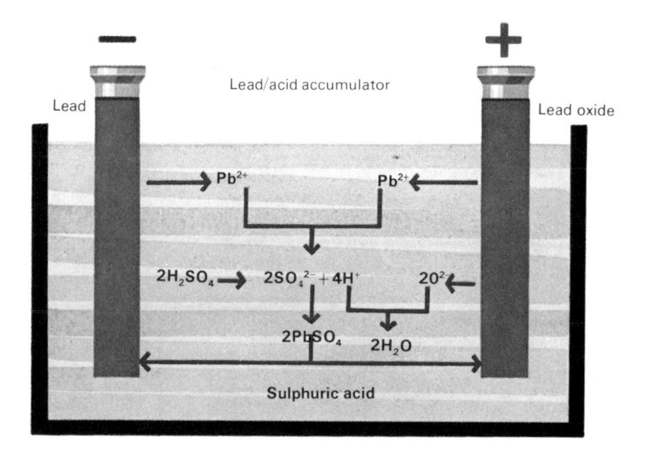

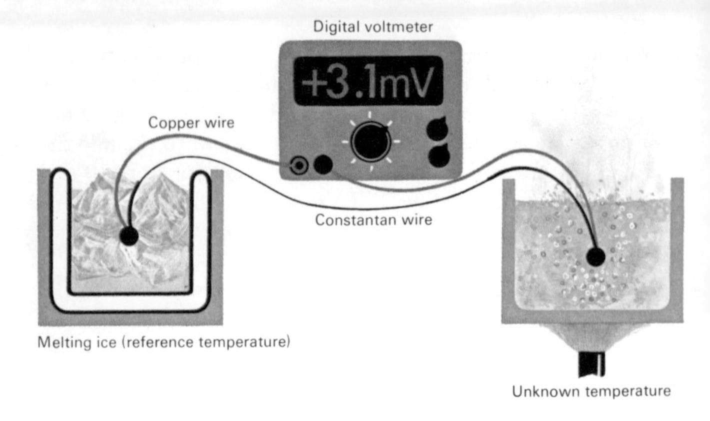

Thermocouple used for temperature measurement

Thermoelectricity

If two different metals are placed in contact, a voltage appears across the junction whose magnitude depends on the temperature of the junction and the nature of the two metals used. If wires of two different metals are connected to form two junctions and the junctions are kept at different temperatures there will be a difference of potential between them and a current will flow. This is called the *Seebeck effect* (after Thomas Seebeck). The voltage across the two junctions gives the temperature difference between them. Such an arrangement is called a *thermocouple*, and is sensitive to rapid

Net radiometer (*left*) measures the difference between solar radiation and radiation reflected from the ground. A Peltier element (*right*) produces a temperature difference.

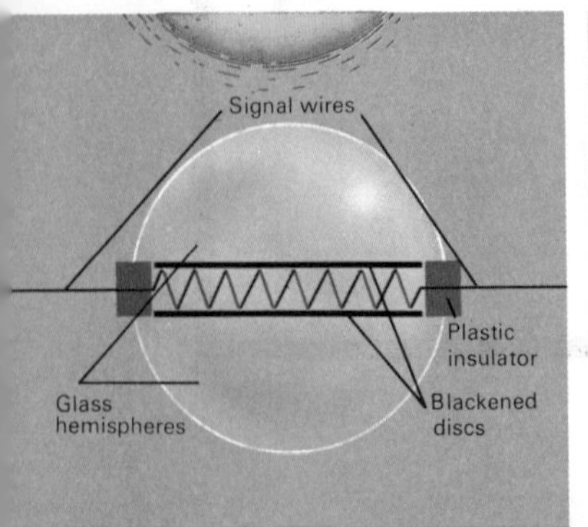

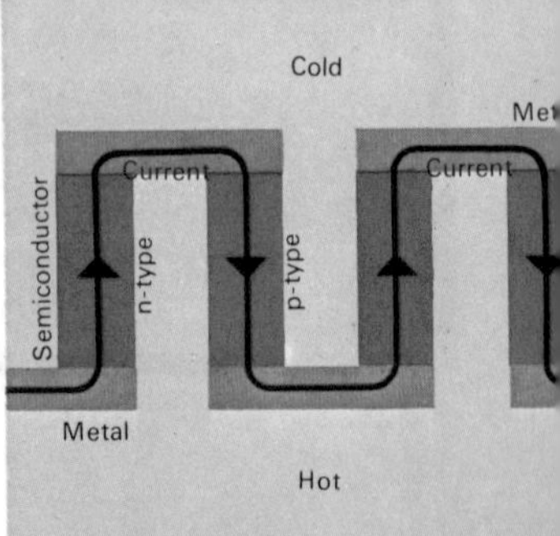

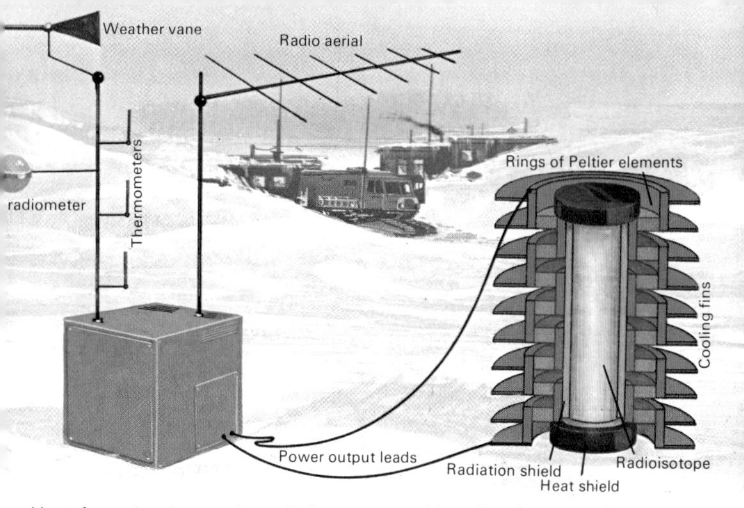

Heat from the decay of a radioisotope provides electric power for an automatic weather station.

temperature fluctuations. The sensitivity is increased by connecting a number of junctions in series.

The *Peltier effect*, discovered by Jean Peltier, is the converse of the Seebeck effect. If a current is passed through a circuit containing junctions of two different metals, heat is liberated at one junction and the other one becomes cooler; if the current is reversed, the hot junction cools and the cold one heats up. The Peltier effect is therefore reversible.

Using two metals as a junction, the Peltier effect is too small to have any practical application. However, certain semi-conductors are up to two hundred times as effective as purely metallic thermocouples, producing temperature differences of up to 75°C. Furthermore, a couple made of metal and a p-type semiconductor will produce a voltage of opposite sign to that of a junction composed of metal and an n-type semiconductor. The arrangement, called a *Peltier element*, will absorb heat from one side and emit it from the other when a current is passed through it. Reversing the current reverses the heat flow. If an object must be kept at a constant temperature, a thermo-couple can be used to operate a switch causing a Peltier element to either heat or cool the object as necessary. Conversely, heating one side of a Peltier element and cooling the other will cause a current to flow.

Radiant energy

Most of the forms of energy discussed so far have relied for their transmission on the atoms and molecules of matter. However, energy can be transmitted through space in the total absence of wires or pipes, or indeed any form of matter. In this form it is called *radiant energy* and it consists of *electromagnetic waves*. It is fairly difficult to imagine waves of energy passing through empty space, but some idea can be gained by thinking of a still pool of water with, say, a matchbox floating at the edge. If a stone is dropped into the centre of the pool, circular waves will radiate from this point to the edges of the pool causing the matchbox to rise up and down as the waves pass. The energy needed to make the matchbox rise comes from the kinetic energy of the falling stone and is transmitted by waves in all directions over the surface of the pool. Electromagnetic waves are in many respects similar to these water waves, with the important difference that they require no material medium for their propagation. Like water waves, they can be thought of as alternating crests and depressions. Waves of this type have four characteristics – *amplitude, wavelength, frequency,* and *velocity of propagation.* The amplitude is the maximum displacement of the wave, the wavelength is the distance between crests, and the frequency is the number of complete cycles occuring in one second.

All electromagnetic radiation travels through space at the same enormous velocity, called the *velocity of light*. If the frequency of the radiation is f (in hertz), it would progress $f\lambda$ metres per second; the velocity therefore is given by: $c=f\lambda$.

Wavelengths cover the enormous range from 10^5 to 10^{-17} metres. At long wavelengths and low frequencies the waves are used for broadcasting radio and television, at higher frequencies they become radiant heat, higher still the retina of the eye is sensitive to them and we call them light. At higher frequencies than light they become ultraviolet, X–, and gamma– radiation. Finally, the highest frequency radiations known to us are those that come from space, called cosmic radiation.

(*Opposite*) The electromagnetic spectrum

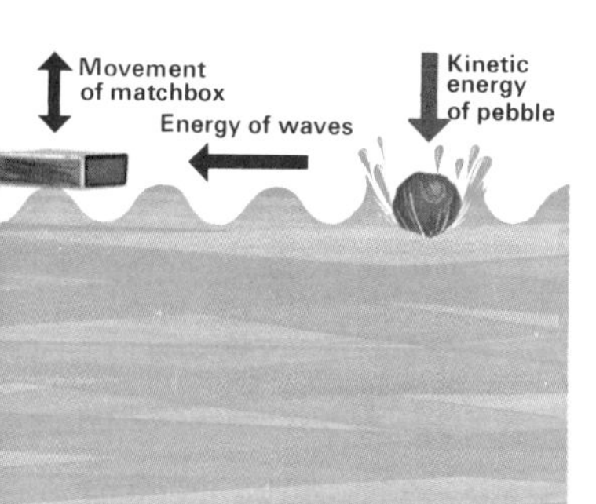

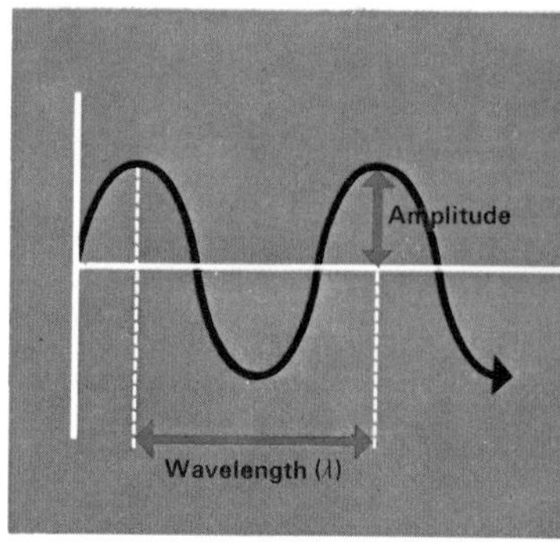

Transmission of energy by waves on water surface (*left*)
Characteristics of travelling wave (*right*)

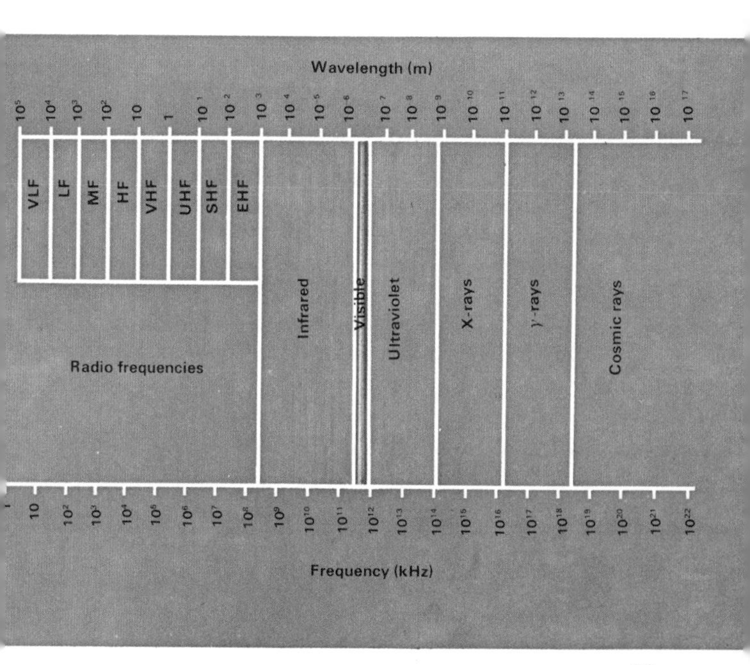

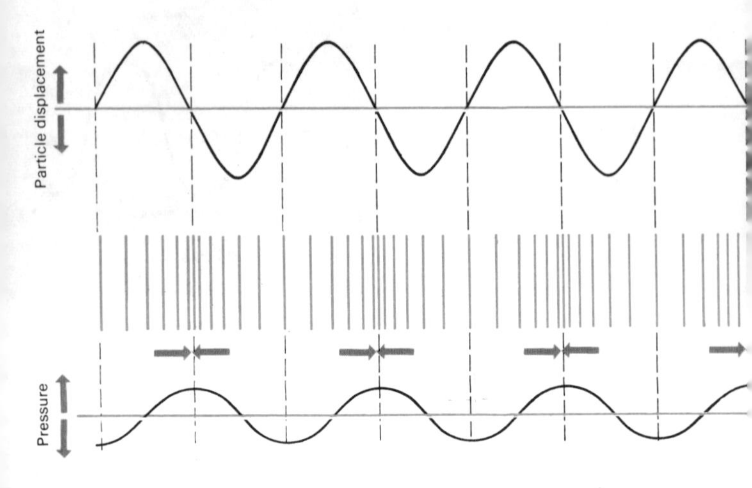

Variation in pressure and position of molecules in a vibrating medium

SOUND

Sound is one form of energy transmitted by waves. The source of sound is a vibrating object in a gaseous, liquid, or solid medium. The to-and-fro vibrations of the source set up an equivalent motion in the molecules of the medium. Each molecule disturbs its neighbour and the energy is transmitted rapidly from one molecule to another in all directions. Unlike electromagnetic waves, sound waves can only travel in a medium. In a vacuum, such as outer space, there can be no transmission of sound.

A sound wave, in which the molecules are moving to-and-fro parallel to the direction of the wave, is called a *longitudinal wave*. Electromagnetic radiation, such as light or ultraviolet, is a *transverse wave* motion. Movement of the molecules in the same direction as the sound wave causes them to be squashed together so that an increase in pressure results. Movement in the opposite direction produces a decrease in pressure. A sound wave can therefore be regarded as a continuous variation in pressure of the medium.

The human ear is a highly specialized organ for collecting sounds and transmitting information about the acoustic environment to the brain. It is able to detect minute pressure variations caused by a sound travelling through air and to convert them to mechanical vibrations of the eardrum, a thin membrane of skin. These vibrations reproduce the original vibrations of the source of sound. A chain of tiny bones – the hammer, anvil and stirrup – connect the eardrum to another much smaller membrane, which transmits the vibrations to the fluid in the cochlea. The cochlea is a shell-shaped organ of about $2\frac{1}{4}$ turns, that converts the mechanical vibrations into electric nerve impulses for transmission to the brain. Inside the cochlea are thousands of tiny hairs with nerve cells at their roots. Movement of the fluid in the cochlea causes the hairs to move, and as the hairs move so nerve impulses are sent to the brain. It is thought that the cochlea can distinguish between different frequencies and convey to the brain information about the loudness of each frequency. The human ear is sensitive to frequencies ranging from 20 – 20 000 Hz (the *audio frequencies*) although the range varies with the individual and falls off with age.

Structure of the human ear

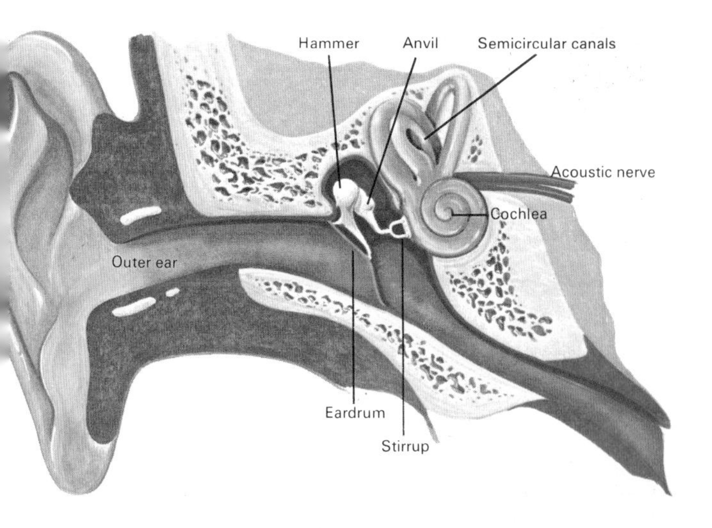

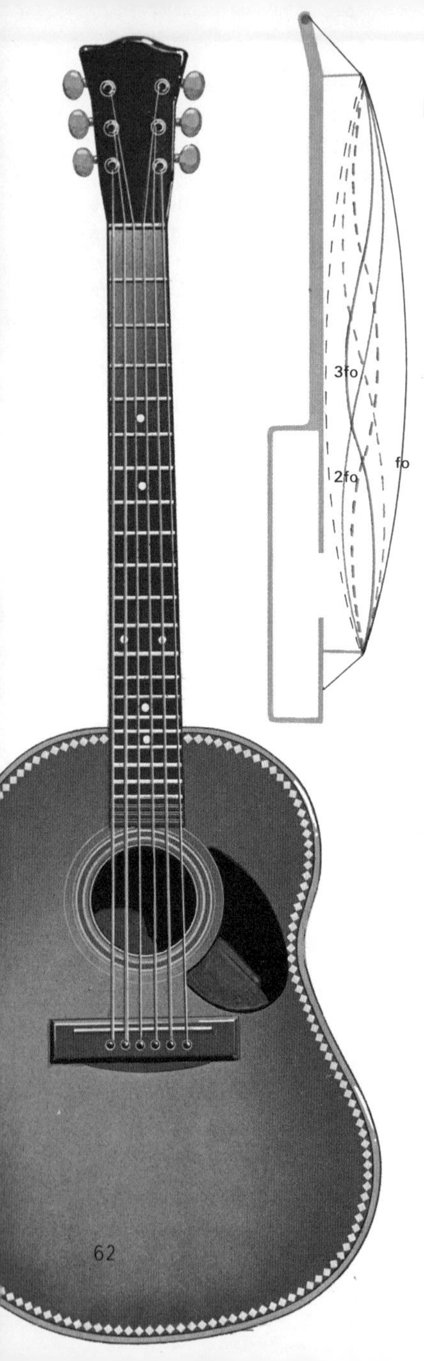

Musical notes

We judge sound as being musical or unmusical by the sensation it produces in the brain. Musical notes have three definite characteristics; *loudness, pitch,* and *quality.*

Loudness is a sensation and therefore depends on the individual. It does, however, increase or decrease as the *intensity* of the sound increases or decreases. The intensity is related to the square of the vertical displacement – the *amplitude* – of the wave. As the energy of the source of sound is dissipated through the medium, the vibrations grow smaller and hence the loudness decreases.

The pitch of a note depends only on its frequency. For stringed instruments, the frequency, f, of the note depends on the length, l, of the string, its tension, T, and its mass per unit length, d; $f=\sqrt{(T/d)}/2l$. In the piano, low frequencies are produced by long thick wires, high frequencies by shorter and thinner wires. A stringed instrument is tuned by tightening or loosening the string. In woodwind and brass instruments the length of the air column inside the instrument determines the frequency.

Holes or stops in the instrument are covered or uncovered to produce different frequencies.

A musical note is not composed of a single frequency but is coloured by the presence of other frequencies. Each particular note has a characteristic frequency called the *fundamental frequency*, f_0, or the *first harmonic*. The other frequencies present are the *upper harmonics* or *overtones* which are whole number multiples, $2f_0$, $3f_0$, etc., of the fundamental. If a thin flexible wire, clamped at both ends, is plucked or struck it is set into vibration and a wave travels along the wire. As the two ends of the wire are unable to move, the wave is reflected in the opposite direction. The waveform associated with the lowest or fundamental frequency is a single closed loop. However the wire can also vibrate at twice this frequency, $2f_0$, producing two loops. The third harmonic is produced by a vibration consisting of three closed loops and so on.

The number of upper harmonics present in a note, and their individual intensities, varies from one instrument to another; notes from a particular instrument therefore

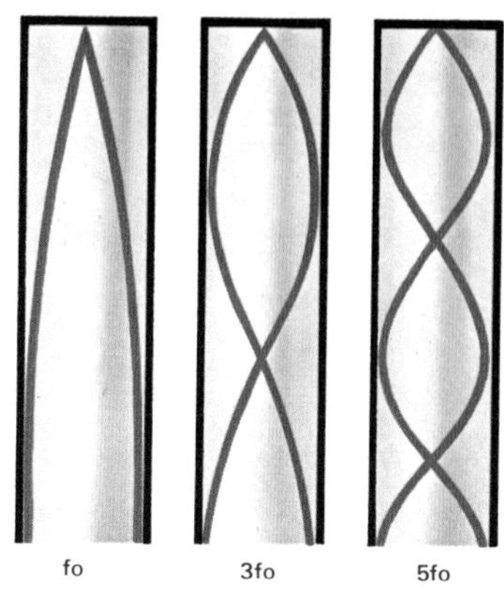

At the open end of a pipe air is free to vibrate and the wave has maximum amplitude. At a closed end air cannot vibrate and amplitude is zero. In a pipe open at both ends all harmonics are present (*below*) but only odd number harmonics are present in a pipe closed at one end (*above*).

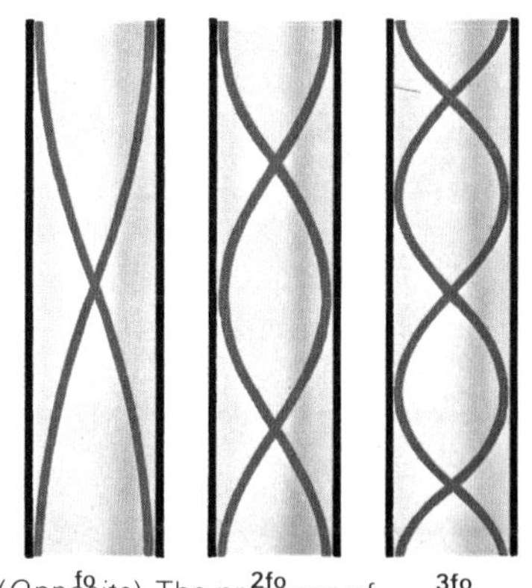

(*Opposite*) The presence of harmonics on top of the fundamental frequency gives colour to a musical note.

have a definite and recognizable *quality*. When a string is plucked, as in a guitar, the intensity of each successive harmonic is much less than its predecessor; when strings are struck, as in a piano, the harmonics are present in equal intensity. The nasal quality of a clarinet is due to a strong third harmonic although the fourth is absent. The waveform is obtained by adding the amplitudes of the component harmonics at each instant of time. This is the principle of *superposition* of waves, and applies to both sound and electromagnetic waves.

There is an enormous range of notes, each having a different frequency, but we select only a limited series of these when we tune an instrument, such as the piano. An acceptable and harmonic musical interval depends not on the frequencies of two notes but on the ratio of their frequencies. Various scales have been used in the past to tune musical instruments. We now use the *equally tempered scale* in which the octave of a note has twice the frequency of that note – it is the second

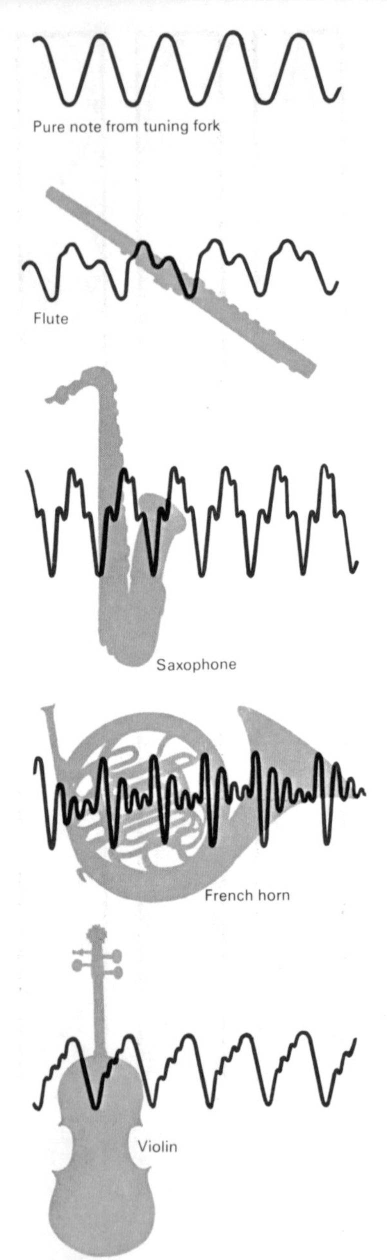

Waveforms of the same note played on different musical instruments

harmonic; each of the twelve adjacent pairs of semitones between the note and its octave have frequencies in the ratio of $\sqrt[12]{2}=1.0595$. The harmonics up to the sixth harmonic are almost exactly the notes of the common chord. For the scale of *a* they are: a', a'', e'', a''', $d^{b'''}$, e''' (the superscript giving the

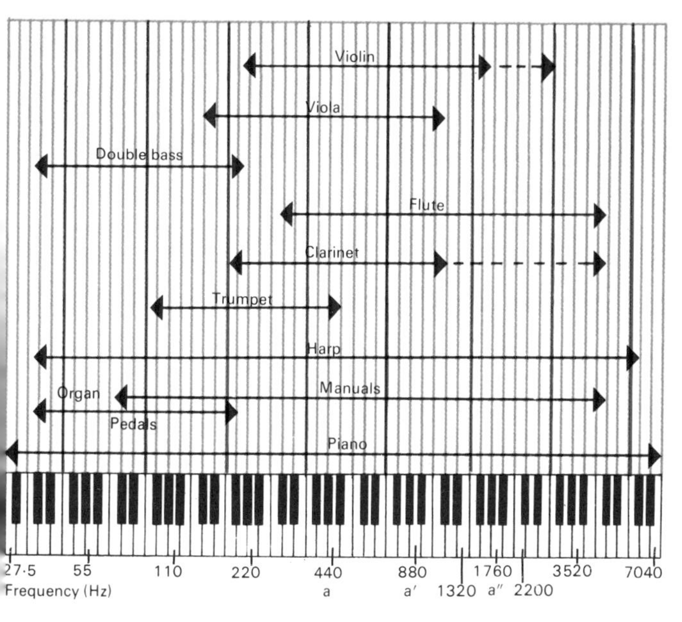

Range of various musical instruments

octave). When these six notes are played simultaneously, they are pleasing to the ear. Some of the upper harmonics, however, such as the seventh and ninth, are not associated with notes of the scale and when they are produced with the other six, they sound most unpleasant. Instruments are therefore designed to reduce these upper harmonics, to a minimum.

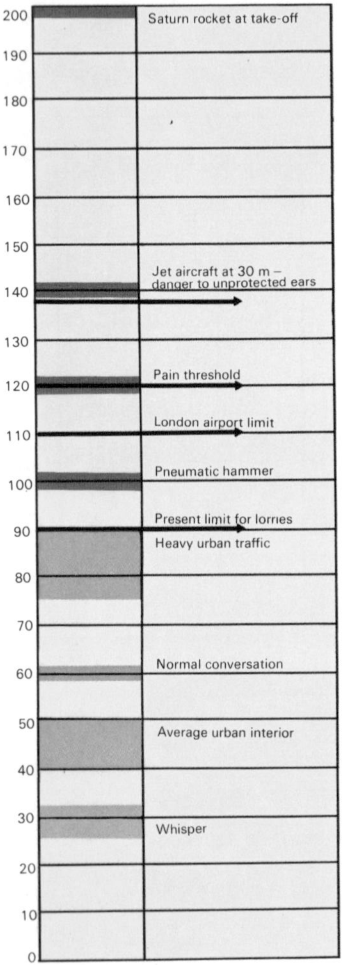

Noise

Noise is any unpleasant sound of irregular frequency that is undesired by the recipient. Our environment is full of noise coming from sources such as air and road traffic, industrial processes, and amplified music, so that legislation has been necessary to control noise levels. The intensity of a sound is measured by comparing its power, P, with the power of a sound, P_0, that is only just audible. As the human response to any stimulus, especially sound and light, is proportional to the logarithm of the stimulus, a logarithmic unit, called the *decibel* (dB) is used: $1\text{dB} = \frac{1}{10} \log_{10} (P/P_0)$.

Usually, a single sound has a constant frequency; for a moving sound, however, such as a train whistling through a station, a drop in pitch is heard as the source of sound passes a stationary observer. This is an example of the *Doppler effect*, which is observed with moving sources of both sound and light. A series of circles, called *wavefronts*, whose centre is the source of sound, indicate the distance a particular sound wave has travelled after a time t, $2t$, $3t$, etc., a wavefront being emitted every t seconds.

Saturn rocket at take-off
Jet aircraft at 30 m – danger to unprotected ears
Pain threshold
London airport limit
Pneumatic hammer
Present limit for lorries
Heavy urban traffic
Normal conversation
Average urban interior
Whisper

200 190 180 170 160 150 140 130 120 110 100 90 80 70 60 50 40 30 20 10 0

Decibel ratings of various sounds

If the source moves, the wavefronts are crowded together in the direction of motion and are spread out in the opposite direction. The distance between wavefronts is proportional to the wavelength. The velocity, v, of sound waves in air is constant (332 m/s) and equals the frequency of the waves times the wavelength. Consequently, as the moving source approaches a stationary observer, the reduced wavelength causes the frequency, f_s of the sound to be higher than when the source is stationary: $f_s = f_o v/(v - v_s)$, where f_o is the stationary frequency. When the source of sound is receding from the observer, the wavefronts are more spread out than usual and the frequency, f_c, will appear lower: $f_c = f_o v/(v + v_s)$. The observed change in pitch occurs at the moment the source passes the observer.

The wavefronts crowd closer and closer as the velocity of the sound source increases. When a supersonic aircraft flies at the speed of sound, the wavefronts coincide and a *sonic boom* is heard by observers on the ground beneath. The boom results from an abrupt drop in pressure that sets up a *shock wave*.

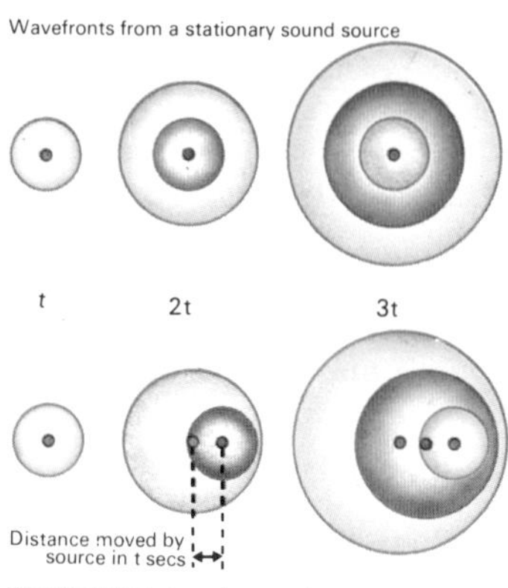

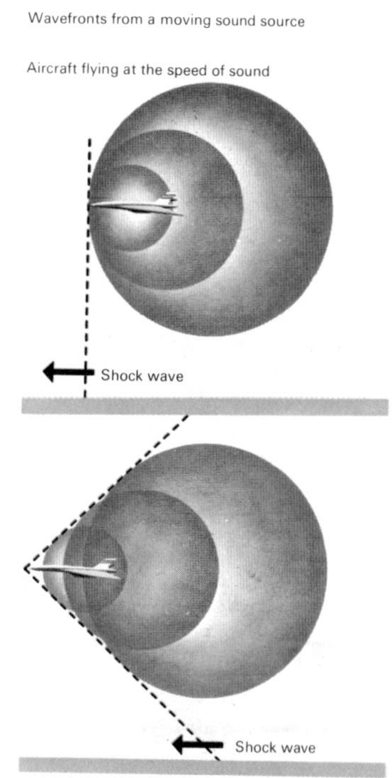

Wavefronts from stationary and moving sources

The conversion of sound into electric currents

Although sound waves are not electromagnetic waves, sound can be transmitted through space using electromagnetic waves. The first step in this process is the conversion of sound waves into an oscillating electric current by means of a microphone. The sound waves travelling through the air fall on a thin metal diaphragm which is thus made to vibrate at the same frequency as the sound waves. Behind this diaphragm is an electromagnet, the magnetic field of which is cut by the vibrating diaphragm. The movements of the diaphragm cause a current to be induced in the coil of the electromagnet, whose oscillations are identical to those of the diaphragm and the sound waves. This oscillating current can be lead through wires to any destination and recreated into a sound wave by an exactly similar process in reverse. This is the principle of the telephone, although sometimes different types of microphone are used.

The telephone has its limitations – to recreate louder sounds an amplifier and loudspeaker are required. Amplifiers are based on electronic valves or transistors. The valve, as its name implies, will only allow current to flow in one direction. The simplest valve used for amplification is the *triode*; this

Transmission of sound by telephone

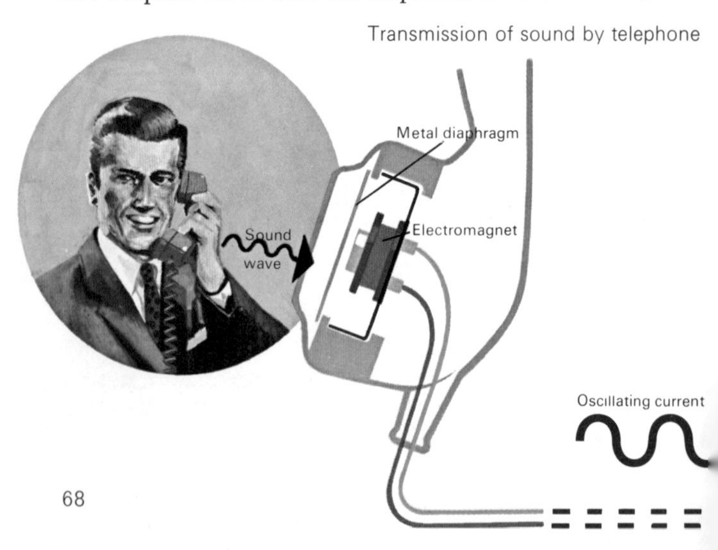

consists of an evacuated glass tube containing three electrodes: a positive *anode*, a negative *cathode*, and between them a *control grid* through which electrons can pass; the cathode is usually heated by a separate filament. When the cathode is connected to the negative terminal of a battery and the anode to the positive terminal, electrons will be able to escape from the heated cathode and will be attracted to the positive anode: a current will flow through the valve. But it will only flow in one direction: if the anode is negative with respect to the cathode, no current flows.

As the electrons have to pass through the grid on their way to the anode, altering the voltage on the grid will alter the number of electrons flowing through the valve. The more positive the grid with respect to the cathode, the more current will flow through the valve; the more negative, the smaller the current. If a weak oscillation is fed between the grid and the cathode, a much stronger oscillation can be obtained from the anode-cathode circuit. The energy output of the anode circuit does not depend on the input to the grid and therefore, by suitable design of the valve and choice of *grid bias* (steady voltage applied to the grid), a very large amplification can be obtained.

Music consists of regular sound waves. In a recording studio these sounds are converted, by a microphone, into

electric currents oscillating at the same frequencies. The output of the microphone is then amplified so that it will make a stylus vibrate (using the same electromagnetic effect as the microphone) in a long spiral groove in a wax master record. The vibrations of the stylus create a series of troughs and depressions in the walls of the groove of the wax record that correspond to the original sound waves. The master record is then electroplated and copies are made of it. These copies are used to stamp thousands of replicas on thermoplastic (vinyl) discs. These are gramophone records.

Extracting the sound from the record depends on the same process in reverse. A jewelled stylus in the pick-up is allowed to follow the groove in the record – the troughs and depressions on the walls of the groove make it vibrate in the same way as the cutting stylus. The pick-up then converts these vibrations into oscillating electric currents. As a typical lightweight pick-up only produces a hundred or so millivolts, it is necessary to use an audio-frequency amplifier in order to operate a loudspeaker. The illustration shows two triodes in series being used as an amplifier. The capacitors C1 and C2 allow only oscillating currents to pass, so they isolate the valve from d.c. voltages in other parts of the circuit. VR is a variable resistance (potentiometer) which acts as a volume

Triode valve and conventional representation

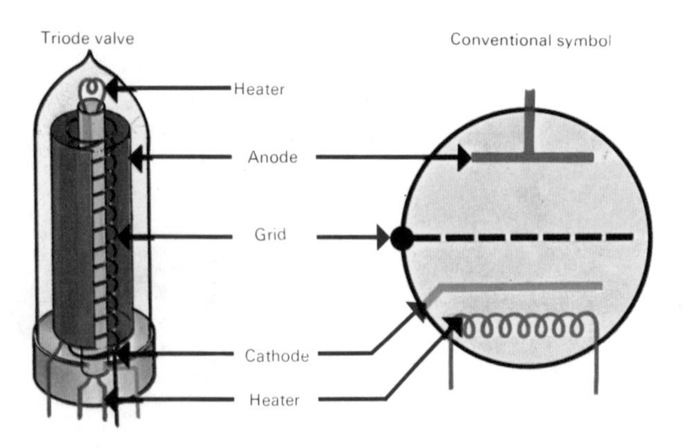

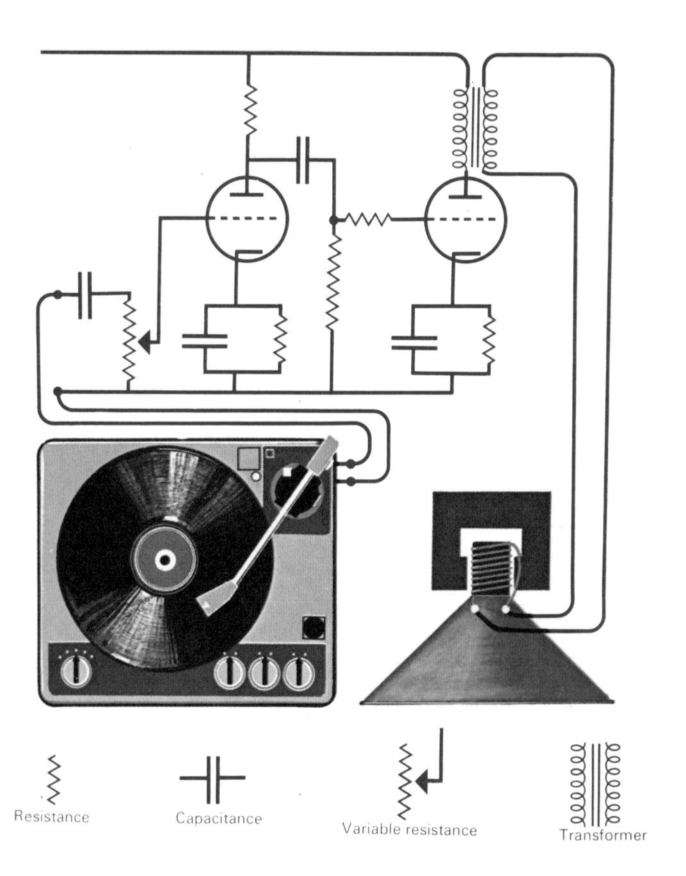

Amplifier circuit of record player

control. R1 is called the anode load resistance and it is through this that the valve current flows. T1 is a transformer coupling to the loudspeaker. The principle of a moving coil loudspeaker is also illustrated: the paper cone is made to vibrate by the interaction of the magnetic field of the permanent magnet and the magnetic field of the speech coil that is connected to the amplifier output. The larger the cone, the more air will be moved and the louder the sound.

Radio transmission of sound

If the output of an audio-frequency amplifier is fed to an aerial, an electromagnetic wave of audio-frequency will be generated. However, the wave would not travel very far as the wavelength would be very long indeed (at 1 kilohertz it would be 3 000 000 metres) and gigantic aerials would be required. Heinrich Hertz (1857–94) discovered that if a radio-frequency oscillation ($10^3 - 10^{11}$ hertz) is fed to an aerial, it will be propagated in all directions for very long distances. These radio frequencies (r.f.) are, of course, very much higher than the audio-frequencies (a.f.) that correspond to sound, however the a.f. and the r.f. oscillations can be combined so that they can both be transmitted together. There are two ways of doing this. In *amplitude modulation* (a.m.) the amplitude of the r.f. wave (or *carrier wave,* as it is called) is modulated by the a.f. In the transmitter the r.f. carrier wave is generated and modulated by the a.f. In the receiver, the modulated carrier is *demodulated* – that is the negative swing of the wave is cut off – and the demodulated carrier is then fed to an audio-frequency amplifier which recognizes the a.f. shape of the carrier but does not respond to the r.f. carrier itself. The amplified a.f. wave is then converted into sound by the loudspeaker.

In *frequency modulation* (f.m.), it is the frequency of the carrier that is modulated by the a.f.

Resonant circuits

Each radio transmitter, whether using a.m. or f.m., has its own characteristic carrier wave frequency. When a radio is tuned to BBC Radio 4 it is being tuned to a carrier wave frequency of 903 kilohertz. But how is the carrier wave produced and how is a radio tuned to select just one particular frequency from all other frequencies being broadcast throughout the world? Both these processes depend on the *resonant circuit*.

The idea of resonance is familiar in acoustics. If an object has a characteristic frequency at which it will vibrate, it can be stimulated into vibration by a sound wave of this frequency. Lay a small piece of paper on the piano strings of C above middle C and then play middle C – the paper will jump

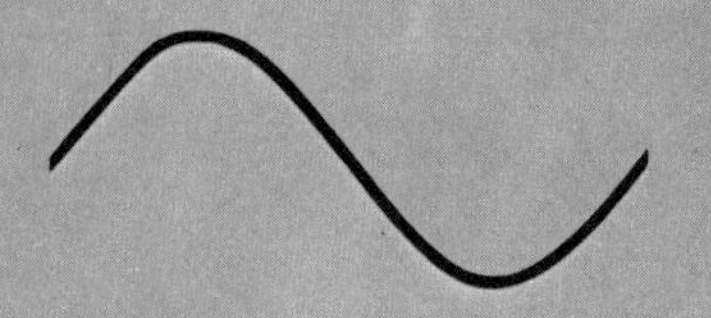

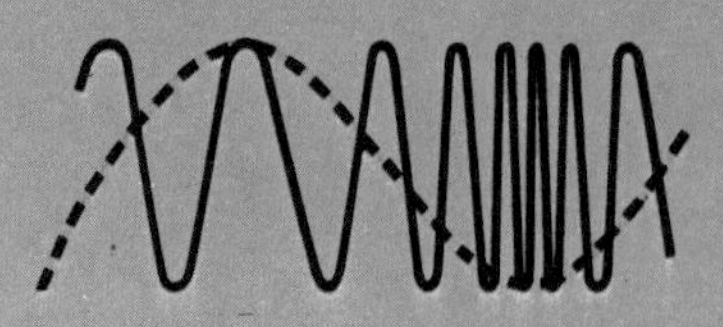

Frequency modulation (*upper*) and amplitude modulation (*lower*)

Carrier wave (radiofrequency)

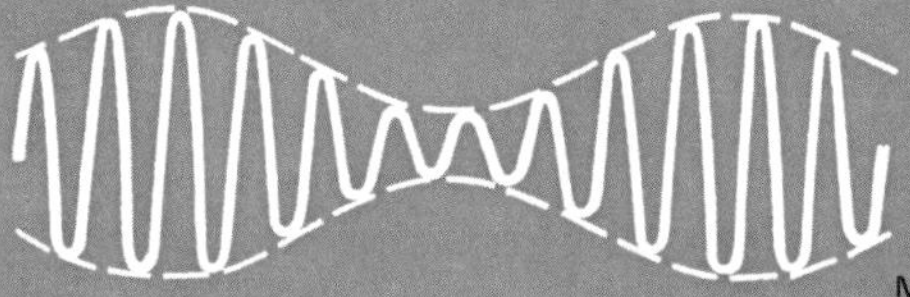

Demodulated carrier wave

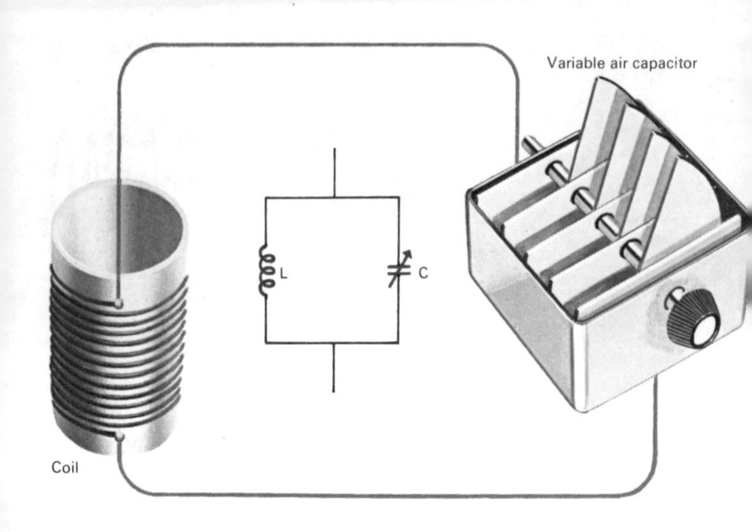
Variable air capacitor
L
C
Coil

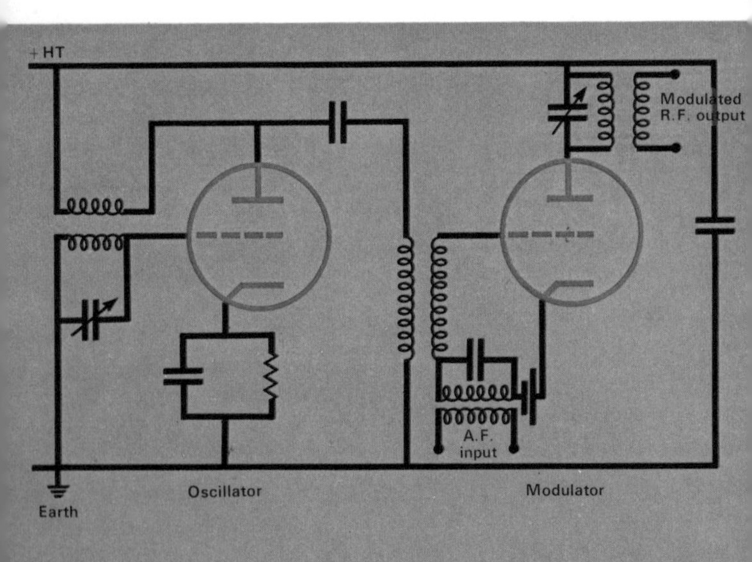
+HT
Modulated
R.F. output
A.F.
input
Earth
Oscillator
Modulator

off because the strings are resonating as a result of stimulation by the sound waves of the second harmonic of middle C. In electronics a similar process occurs in a resonant circuit at radio frequencies, the frequency of maximum response being the *resonant frequency*. A resonant circuit can consist of an induction coil and a capacitor in parallel or in series. In the parallel circuit, when the capacitor is charged up and then allowed to discharge through the induction coil, a magnetic field will be created round the coil which will collapse after the capacitor is completely discharged. This changing magnetic field will induce a secondary voltage in the coil in the opposite direction to the first. This secondary voltage will then charge up the capacitor in the opposite direction to its previous charge. When the capacitor discharges itself through the coil again one cycle of the oscillation will have been completed; this oscillation will continue until the resistance of the circuit has consumed all the energy of the initial charge. The frequency of oscillation will depend on the inductance, L, of the coil and the capacitance, C, of the capacitor; at the resonant frequency, f, $f = 1/(2\pi\sqrt{LC})$; when f is in hertz, C is in farads and L in henries.

The resonant circuit will not oscillate indefinitely unless energy can be supplied to it to overcome the resistance of the flow of electrons back and forth through the wires. This energy can be supplied by using a triode valve. If the oscillating current of a resonant circuit is fed to the grid of a triode, an identical but stronger oscillation will be produced in the anode circuit. If this is fed back into the resonant circuit through a second induction coil linked to the first, the device will oscillate indefinitely as energy is fed into the circuit from the source that operates the valve. This is the basis of a simple transmitter. V1 in the illustration is a triode valve being used as an oscillator (a tuned grid circuit) and V2 is a triode being used as a modulator (grid modulation). In this simple device, the modulated output would be amplified before being fed to an aerial.

(*Top*) A resonant circuit

(*Bottom*) A simple transmitter

Propagation of radio waves

Radio waves are propagated through the earth's atmosphere in one of two ways. *Ground waves* travel over the earth's surface, and as they travel in straight lines this means of communication is only possible over relatively short distances. *Indirect* or *sky waves*, however, enable radio communication to be maintained between places that are not in a direct line. This means of communication depends on a layer of ionized gas in the upper atmosphere called the *ionosphere* – the ionosphere reflects these indirect waves back to earth.

The ionization of the atoms and molecules in the ionosphere is largely caused by the action of ultraviolet and X-radiation from the sun, and therefore conditions vary from day to night. At night, the ionization in the lower part of the ionsphere called the *Heaviside layer* (or E-region – about 90–150

The propagation of radio waves

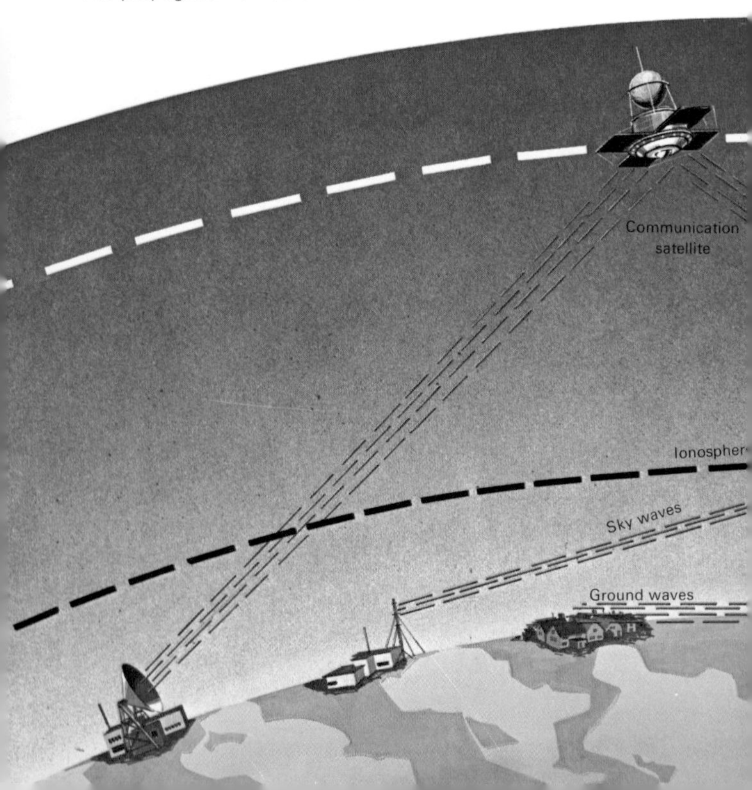

km above the earth) falls off as in the absence of sunlight the ions and electrons tend to recombine. However, the higher layer called the *Appleton layer* (or F– region – 150–400 km) remains ionized longer, as it is less dense and there are consequently fewer collisions between ions and electrons.

The very high frequencies (VHF) and ultra high frequencies (UHF) used in television broadcasting penetrate both layers of the ionosphere with little or no reflection – it is for this reason that TV programmes cannot be broadcast over long distances. However, this problem is overcome by *communication satellites* in *stationary orbits*. A stationary orbit is one in which the satellite rotates in the plane of the equator at exactly the same speed as the earth, so that it remains in exactly the same place with respect to the earth. A satellite in stationary orbit over mid-Atlantic can be used to receive a signal from America and retransmit it to Europe.

Receivers

A radio receiver has to perform two essential functions. First it has to select the particular carrier wave that is required from all the other radiations over the earth's surface. Secondly, it has to demodulate the r.f. carrier so that the a.f. component can be extracted.

In an a.m. receiver a resonant circuit is used to make the selection. If the resonant circuit of the receiver is tuned to oscillate at exactly the frequency of the required carrier wave, this carrier and this one only will be reinforced – all the other radiations picked up by the aerial will be ignored. In most radios the resonant circuit is tuned by altering the value of a variable air capacitor.

The next step is to demodulate the carrier. Once again a triode can be used. If the modulated carrier from the resonant circuit is fed into the grid of the valve, and the bias is correctly arranged, only the positive half of the swing of the carrier wave will be passed through the valve, the bottom or negative half of the swing will be cut off. This is called *rectification*. The output of the anode circuit of the valve can then be fed to headphones, or after amplification, to a loudspeaker.

Television

A radio signal can be transmitted as a frequency modulated wave. The same technique is used to transmit picture information in television – the output of a television camera is fed to a transmitter and broadcast.

The simplest form of TV camera is called the *vidicon,* in which the camera tube is only about 15 cm long and 2.5 cm diameter. The optical system focuses an image onto a transparent conducting plate, behind which is a layer of antimony trisulphide. This is a *photoconductor* – its resistance decreases when light falls on it. An electron beam scans the back of the photoconducting layer, and when it passes an illuminated area, a voltage signal is picked up from the conducting plate.

The electron beam is made to scan the plate in 625 horizontal lines, thus making one *frame*; this happens 30 times a second. Each frame is built up by first scanning the odd-numbered lines and then the even. At the end of each line a short negative voltage pulse is added, with a longer one at the

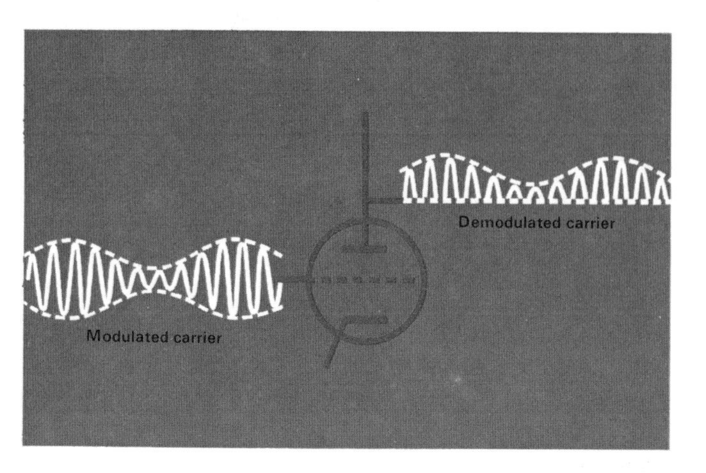

(*Upper*) A carrier wave is demodulated by cutting off the negative voltage swing. (*Lower*) A simple receiver

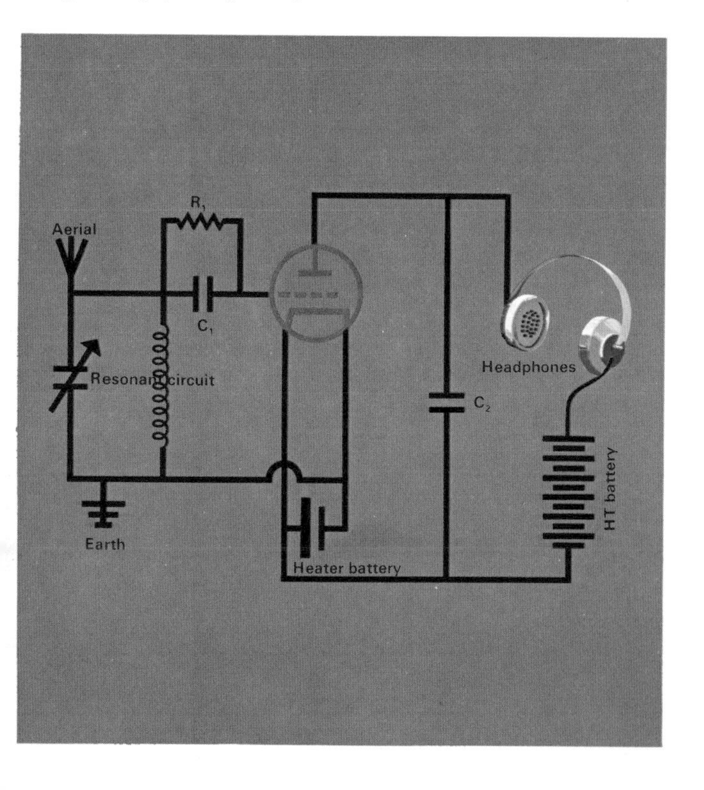

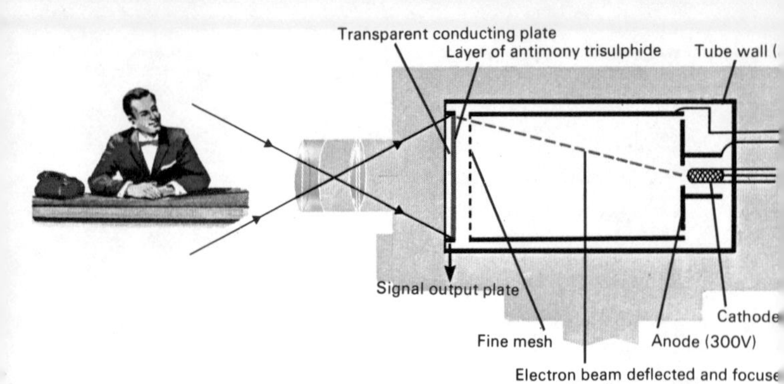

end of each frame. These lines are at a slight angle because the beam is moving steadily down the screen while it is scanning.

In the receiver, the incoming signal is used to vary the intensity of an electron beam scanning the screen of a *cathode ray tube*. This device consists of an *electron gun* for producing the beam, focusing electrodes, and magnetic coils around the neck of the tube for deflecting the beam so that it scans the screen. The inside of the screen is coated with zinc sulphide, which glows white when struck by electrons. The screen is scanned in exactly the same way as the plate in the camera, with the end-of-line and end-of-frame pulses keeping the camera and receiver in step.

In the vidicon camera (*upper*) an electron beam scans a photo-conducting plate. The pulses generated by the plate are transmitted and used to construct the image on a cathode ray tube face (*lower*). In the cathode ray tube (*right*) an electron beam scans the tube face synchronously with the camera beam. The intensity of the beam is controlled by the transmitted signals.

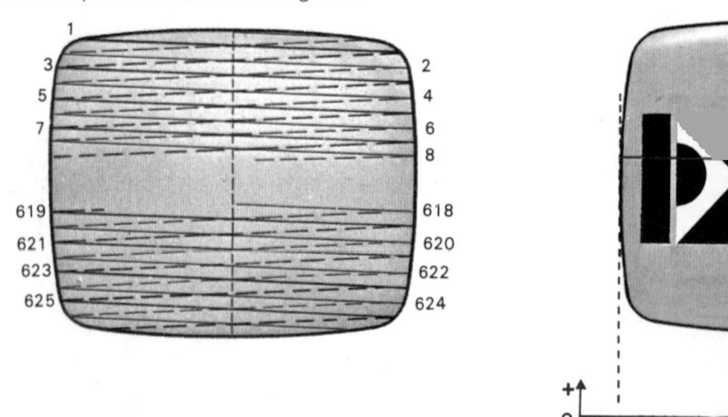

Radar

Radar is an acronym for *ra*dio *d*etection *a*nd *r*anging. It was developed during the second world war to enable enemy aircraft to be detected. It is based on the directional beaming, by a radio antenna, of a series of pulses of radio waves of very short wavelength towards a distant object; this beam is very narrow and can travel a greater distance in the earth's atmosphere than light, as it is not scattered by small dust particles. Some of these pulses are reflected back to the point of origin, and are received by the same antenna, which is switched electronically from transmitting to receiving immediately after each pulse. The direction of the target is determined from the direction of the reflected beam of pulses; the range (or distance) of the target is determined by the delay between sending and receiving a pulse, the echo of one pulse being received before the next pulse is emitted. If an echo is received after a time, t, (a few microseconds) the distance between antenna and object is $\frac{1}{2}ct$, where c is the velocity of light. In some systems, the velocity of the target can be determined from the frequency shift of the echo, due to the Doppler effect.

The radar signal is displayed on the fluorescent screen of a cathode ray oscilloscope. This tube is either an *A-scope* in which distance is represented by a horizontal line and echo strength vertically, or a *plan positive indicator* (PPI), in which the target is represented by a blip on a radial line whose

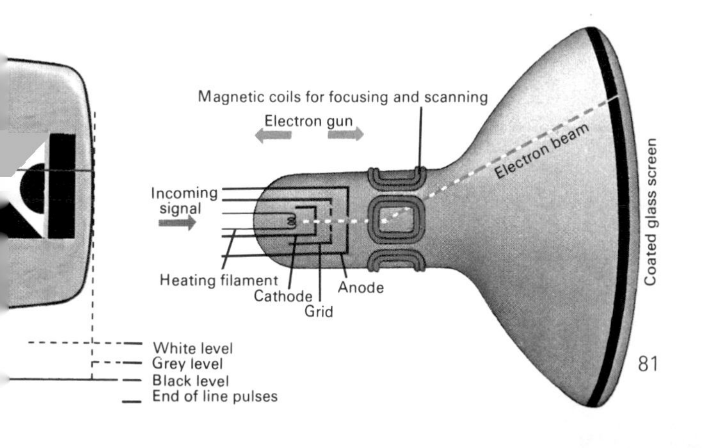

direction corresponds to the direction of the rotating antenna.

There are three main requirements of a radar set: high frequency (usually between 200 MHz and 35 GHz) to give good reflection from small targets; high power, since only a very small fraction of the transmitted beam returns; good directionality, to pin-point the target.

The frequency and power for a radar transmitter cannot be provided by conventional valves; special devices, such as the *magnetron,* are used. The magnetron is essentially a diode placed between the poles of an electromagnet. It produces its own high frequency oscillations by the interaction of electric and magnetic fields.

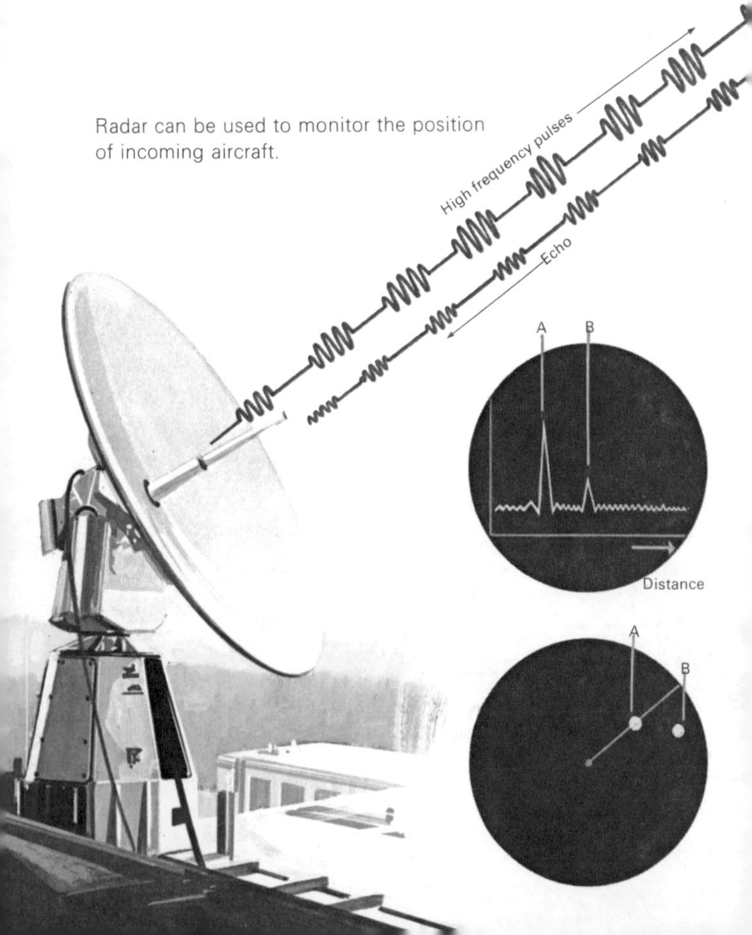

Radar can be used to monitor the position of incoming aircraft.

Semiconductors

All electronic circuits and devices described so far have depended on valves. However, since the 1950s there has been a steady trend towards replacing valves with transistors, chiefly because they are smaller, lighter, less fragile, require no heater current, and work at much lower voltages. Transistors are *semiconductor* devices.

Most metals are good conductors of electricity because they contain large numbers of free electrons. Most non-metals have very few electrons for conduction, as their electrons are tightly bound to their respective nuclei. These substances are *insulators*. There is a third class of materials in which some of the electrons have sufficient energy to break free from their bound positions – these are semiconductors; silicon, germanium, and selenium are examples. The number of free electrons in a semiconductor is reduced by decreasing its temperature: at absolute zero an ideal semiconductor would have no free electrons. The higher the temperature, the greater the energy of the electrons, and the greater the percentage that are able to break free from their bonds. The resistance of semiconductors therefore decreases with temperature whereas it increases with metals.

In a pure semiconductor crystal the conductivity is small. However, if certain impurities are added to the crystal, the conductivity is increased considerably. A pure germanium crystal has four electrons in its outer valency shell, each of which is bound by a covalent bond to a neighbouring atom. An arsenic atom contains five electrons in its outer shell. If

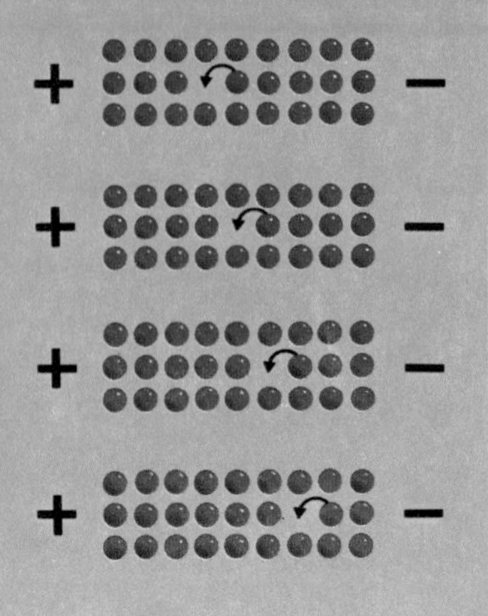

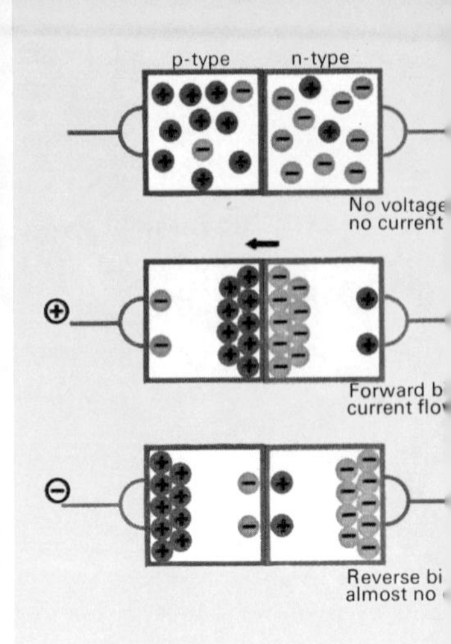

(*Upper left*) Movement of positively charged 'hole' in an electric field
(*Upper right*) Rectification at a p–n junction
(*Lower*) Germanium lattice with arsenic impurities

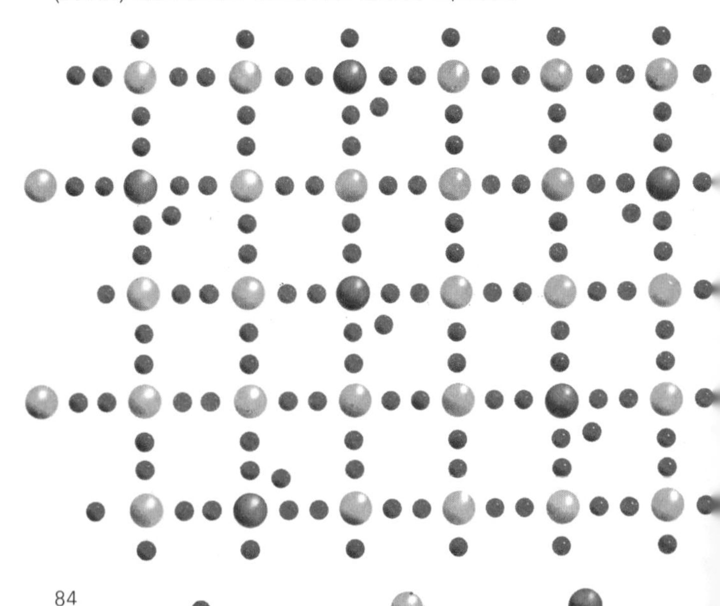

a certain concentration of arsenic is added to germanium during crystallization, there will be a spare electron for every arsenic atom in the lattice. These electrons are available for conduction of a current; this type of conduction is called *n-type*. If, however, atoms (such as indium) with only three outer electrons are added to germanium, *holes* will occur in the lattice. Although valency electrons are not free to drift around, they can fall into an adjacent hole. Holes thus behave as if they were positive charges and a hole current is called *p-type* conductivity. Electrons and holes are both *charge carriers* and in each type of conductivity a few charge carriers of the opposite type will exist. For this reason, holes in p-type, and electrons in n-type, are called *majority carriers*.

If a p-type semiconductor is joined to an n-type, and a voltage applied across the junction so that the p-type end is positive, all the charge carriers will be forced by the field towards the junction, where they combine. Further charge carriers will be formed behind them and the continuous process constitutes a flow of current. Reversing the voltage, however, causes the charges to separate and very little current flows. The semiconductor junction therefore acts as a rectifier.

Transistors

In order that semiconductor devices may be manufactured with exactly the desired characteristics, large single crystals must be grown about 15 cm long, 2.5 cm in diameter, and 99.999 999 9 per cent pure. Small amounts of impurity elements are then added to convert the crystal to the n– or p–type state; this process is called *doping*. By diffusing doping agents into different areas of a semiconducting crystal, it is possible to construct devices having a number of semiconducting junctions.

The *transistor* has two junctions, arranged in an n–p–n or p–n–p order. The central portion is called the *base,* and the outer parts the *emitter* and *collector,* the majority carriers originating in the emitter. The collector is always the larger so that charge carriers can diverge through the base to the collector. The function of the base is to control current flow in a somewhat similar manner to the grid of a valve: small

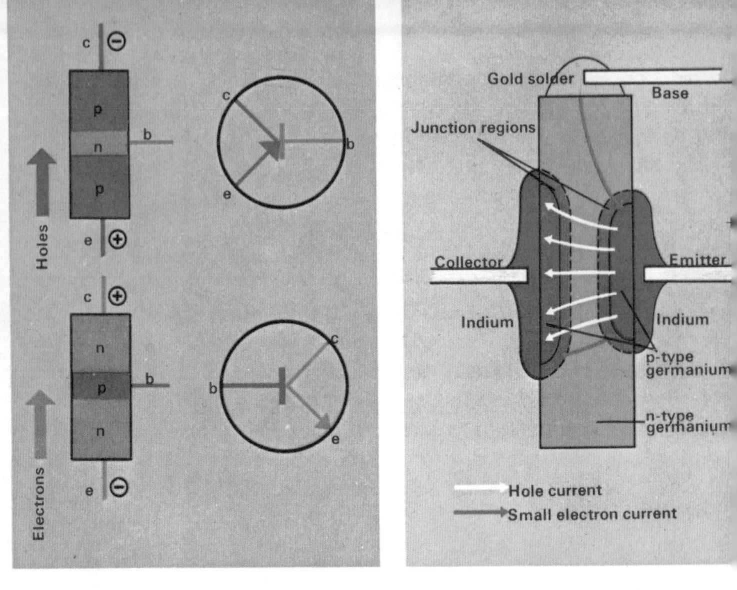

(*Far left*) Transistor types and conventional symbols. (*Left*) Alloy junction transistor. (*Right*) Field effect transistor. (*Far right*) Planar silicon transistor

variations in the current between emitter and base cause large alterations in the current from emitter to collector. The transistor can therefore act as an amplifier and is now used for this purpose in all electronic devices unless valves or other specialized devices are required. There is no difference in principle between n–p–n and p–n–p transistors, but the availability of these two types allows new departures in circuit design.

There are two other important types of transistors besides the alloy junction type. In the thin film *field effect transistor* (FET) the controlling electrode is called the *gate*; it is insulated from the *source* and *drain*, names given to the emitter and collector in this type of transistor. Changes in the electric field around the gate control the current flow, whilst using little or no input current. The FET is made by evaporating materials onto glass, with cutout masks defining the areas where each layer must go.

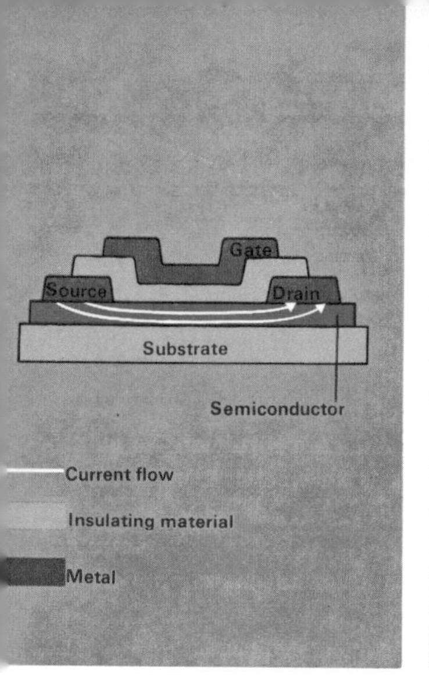

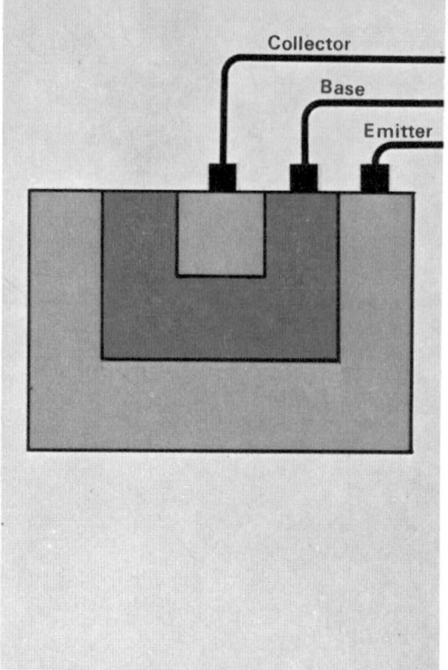

In the *planar silicon transistor*, a substrate of doped silicon is used, and doping agents of opposite type are diffused into the crystal forming n– and p–type regions to which leads are attached. The details of this process, called *diffusion*, are given in the next section: a layer of silicon dioxide is formed on the crystal, gaps are etched in it using a photographic process, and the surface is then exposed to vaporized doping agents.

Integrated Circuits

Since 1955, transistors and other circuit elements have been greatly reduced in size. These elements are now built into blocks in which they are completely inseparable; these are *integrated circuits*. Over one thousand circuits, each containing 20 or 30 components, can be made on a silicon wafer 2.5 cm in diameter and less than 0.2 mm thick. These circuits are rugged, long-lasting, cheap, and are capable of performing electronic functions at extremely high speed due to the proximity of the components.

A succession of masking and diffusion steps is involved in the manufacturing process, each mask containing a specific number of accurately shaped holes. The n–type impurities are diffused into the p–type silicon wafer, and the subsequent layer of silicon dioxide allows a layer of p–type impurities to be diffused only into certain areas. Using different masks, other p–type and n–type diffusions occur in specific regions. The whole *chip* is finally coated with metal, which is then etched to form connections between the devices making up the circuit. The original p–type silicon layer separates the components to minimize interference between them. Two or more superimposed diffusions of n–type (or p–type) impurities produce different depths and concentrations of doping agent. The silicon dioxide mask is an electrical insulator and forms the dielectric in the capacitor.

Integrated circuits are extremely important in computers; the speed at which computers function ultimately depends on

Stages in the manufacture of an integrated circuit

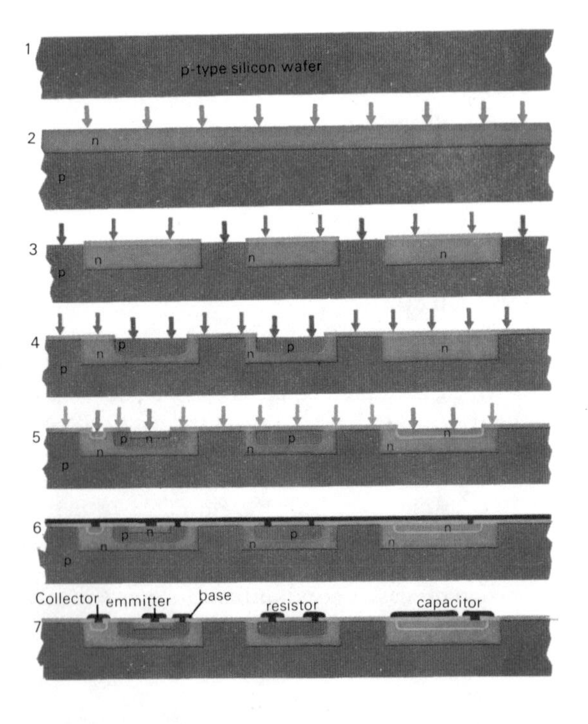

the speed at which the electrons move. The close proximity of integrated circuit elements optimizes this factor. *Digital computers* count in ones and zeros only (the binary notation); these two numbers can be represented by the presence or absence of a current. Integrated circuits in computers are called *micro-logic elements* and are used to pass a particular current, punch a hole, etc.

The simplest micro-logic elements in computers are the *AND* circuit and the *OR* circuit. The AND circuit has an output only if *all* the inputs receive signals, the OR circuit has an output if *one or more* of the inputs receive signals. More complex logic circuits can be built from these two basic circuits.

Infrared radiation

The infrared (IR) region of the electromagnetic spectrum lies beyond the radio and microwave region, and extends to the visible section, the wavelength range being approximately

(*Left*) Integrated circuit with circuit diagram
(*Right*) Micro-logic circuits and (*bottom*) integrated circuit element shown on 1p coin

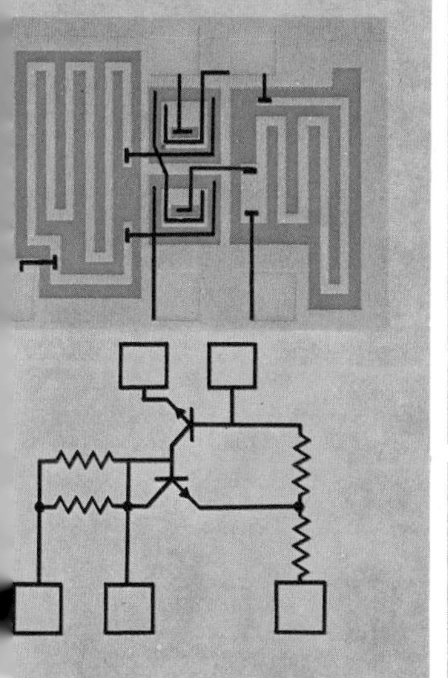

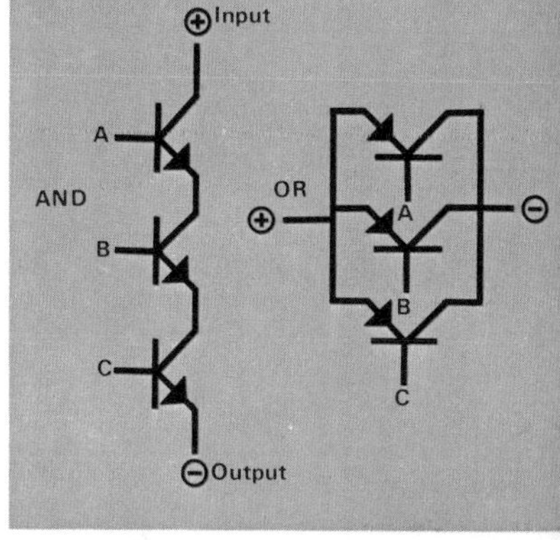

400 – 0.76 micrometres (μm). IR radiation has the longest wavelengths to which the human body can respond directly, the sensation being one of heat. IR is therefore sometimes called radiant heat. All bodies that absorb IR radiation become warmer. An absorbing body will also emit IR; after a time a state of equilibrium is reached in which the total energy absorbed equals the total energy emitted, the emitted radiation having longer wavelengths.

The sun is an important source of IR, emitting wavelengths down to 10 μm as well as visible and ultraviolet radiation.

Comparison of solar radiation absorbed by the Earth with radiation emitted by the Earth

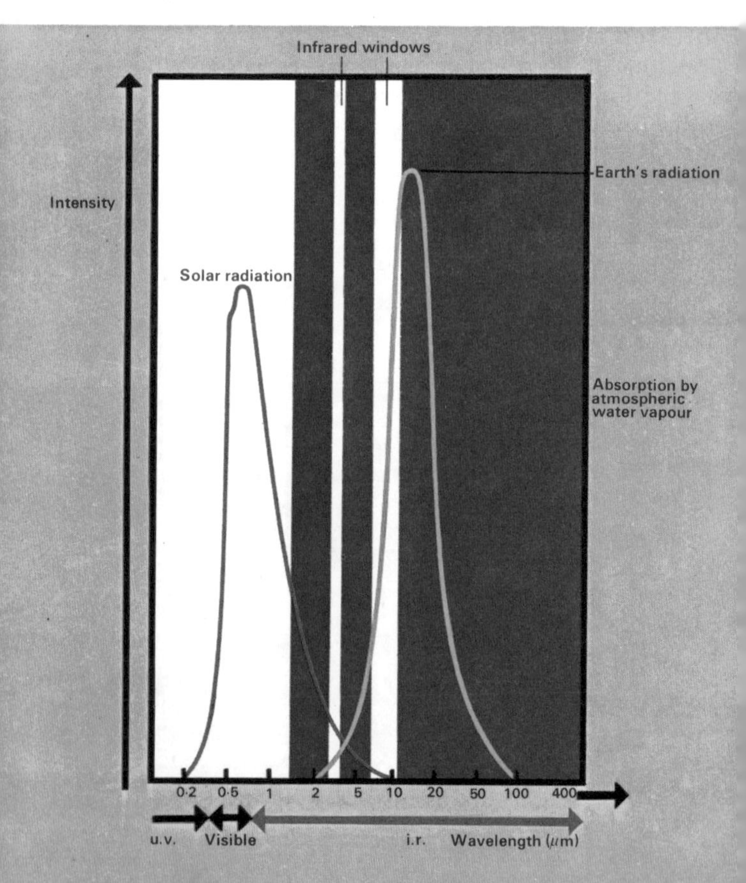

The earth re-emits IR radiation having a wavelength range of about 6.8 to 100 μm. Water vapour in the atmosphere absorbs a large amount of this re-emitted radiation, except for a band between 8 and 11 μm, called the *infrared window*. The fraction of IR emitted by the earth that escapes through this window helps to balance the energy absorbed from incoming IR. This preserves the *earth's energy balance,* maintaining a reasonably constant air and ground temperature.

Electromagnetic radiation is produced by the motion of electric charges and their associated magnetic fields. When the moving charges are vibrating atoms and molecules, the emission is IR radiation. Absorption of IR produces an increase in vibrational speed causing a rise in temperature. As the temperature of the body rises, the frequency of the emitted radiation increases until, eventually, the body glows red-hot, emitting light from the visible region of the spectrum.

A *black body* is a theoretical concept of an ideal body that absorbs all incident radiation without reflection, re-emitting the same energy at lower frequencies. (Both the sun and the earth behave almost like black bodies.)

Before 1900, it was assumed that any amount of energy could be exchanged between radiation and matter. Investigation of black-body radiation proved this to be wrong. Wave theory correctly predicts that the total energy emitted by a black body is proportional to the fourth power of its absolute temperature. However the energy emitted at a particular frequency could not be predicted using a single formula throughout the whole range of IR wavelengths. Max Planck successfully resolved this problem by postulating, in 1900, that the exchange of energy between radiation and matter is not continuous but is made in discrete amounts called *quanta*. The energy, E, of each quantum depends on the frequency, f, of the radiation, $E=hf$, where h, the *Planck constant,* has the value 6.62×10^{-34} joule seconds. The implication of this theory is that atoms and molecules can only vibrate at certain or absorbed, this speed jumps from one permissable value to another. Planck's equation has proved accurate over the entire electromagnetic spectrum. All energy is quantized in this way, but the effects are only significant at the atomic level. Planck's postulate is the basis of *quantum theory*.

LIGHT

The human eye is only sensitive to a narrow range of electromagnetic radiation between the wavelengths 760×10^{-9}m (760 nm) and 380 nm. These are the wavelengths of *light*. The response of the eye depends on the *wavelength,* each wavelength producing the sensation of a different *colour*.

A beam of light can be thought of as a group of single rays all travelling in straight lines in more or less the same direction. The sharpness of shadows and the formation of eclipses are evidence of this rectilinear propagation.

We see an object either by the light it emits, as with the sun or an electric light, or by the light that it reflects, as with the moon. When a beam of light falls on an object some of it is *reflected* from the surface, some may be *transmitted* through the object, and the rest is *absorbed* by it. The *law of reflection* states that the angle of incidence, i, is equal to the angle of reflection, r. A highly polished regular surface, like a mirror, reflects light in a definite direction; a rough surface reflects it in all directions.

The velocity of light in air is close to its maximum velocity of 3×10^8 m/s in a vacuum. However, when a beam of light travelling through air enters another medium, such as glass or water, the velocity is reduced. As the frequency is

(*Below*) Eclipse of the sun viewed from different points on the Earth's surface. (*Right*) Reflection and refraction of light

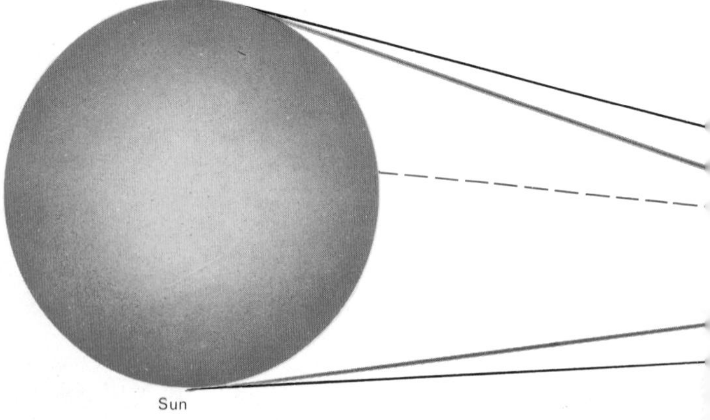

Sun

constant the wavelength is reduced and the direction of the beam is consequently altered. Whenever light passes to a denser medium, at an oblique angle to the surface, it is bent or *refracted* towards an imaginary line perpendicular to the surface at the point of incidence. The ratio of the velocity of light in a vacuum (or in air to a close proximation) to that in a denser medium is a constant known as the *refractive index*, μ. For glass it is approximately 1.5, and for water it is about 1.34. The *law of refraction* states that the refractive index equals the ratio of the sine of the angle of incidence, i, to the sine of the angle of refraction, n.

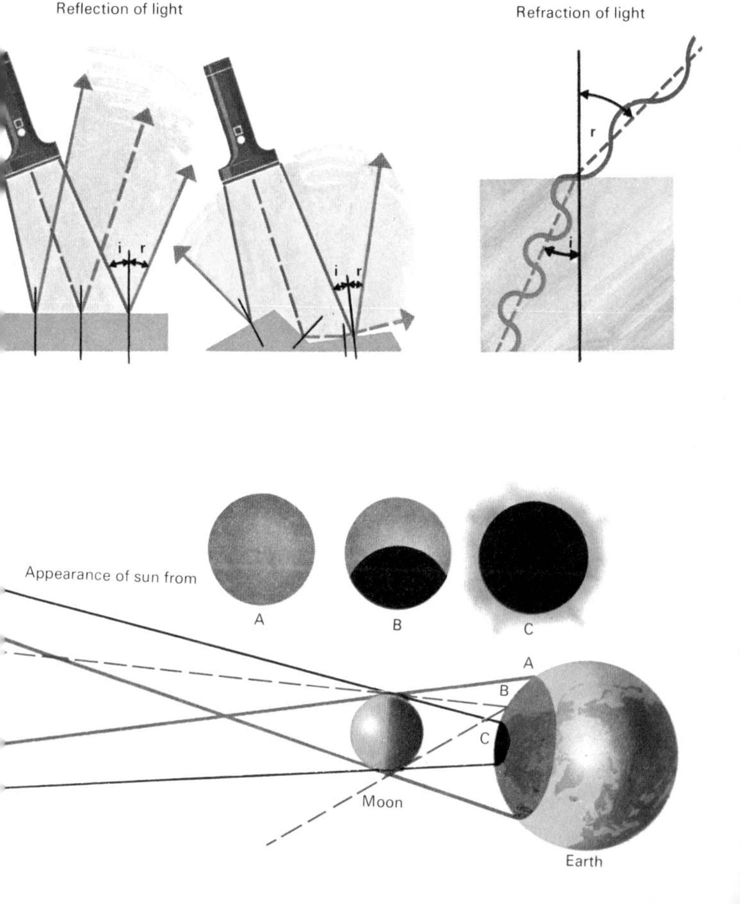

Geometric optics

Light can be bent in any required direction by using curved lenses or mirrors of the appropriate shape. A *convex* (curving outwards) lens or a *concave* (curving inwards) mirror converges a beam of parallel rays to a point known as the *focus*. Concave lenses or convex mirrors diverge a parallel beam so that it appears to come from the focus; the greater the curvature of the surface, the nearer the focus is to the lens or mirror.

A lens or mirror forms an *image* of the luminous or light-reflecting object from which the incident light originates. Light is reflected from a mirror according to the laws of reflection. If the ray is perpendicular to the surface it returns along the same path. Passage of light through a lens follows the laws of refraction. If a ray passes through the centre of the lens no refraction occurs. Using these rules it is possible to determine geometrically, the position of the final image at the point of intersection of these rays.

An image that can be focused on a screen is a *real* image. For some positions of the object, relative to the focus, the rays

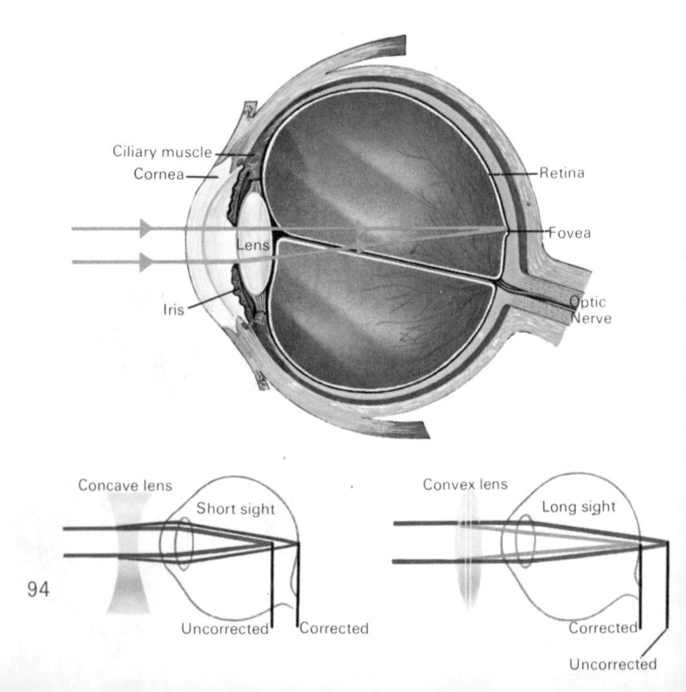

diverge after reflection or refraction and only appear to come from the image, which is then said to be *virtual*. (The image formed by a plane mirror is virtual.) The lens or mirror may also form the image upside-down or *inverted*.

In the eye incident light is refracted both by the *cornea* and by the *lens*, and is focused on a sensitive area of the *retina* called the *fovea*. The retina contains two kinds of light-sensitive cells, *cones*, concentrated in the fovea, and *rods*. These cells send impulses to the brain through the fibres of the optic nerve.

The converging power of the eye lens is controlled by the *ciliary muscles* attached by ligaments to the lens. The amount of light entering the eye can be varied by the *pupil* by increasing or decreasing its diameter. For night vision the pupil is wide open; for bright light it is very small.

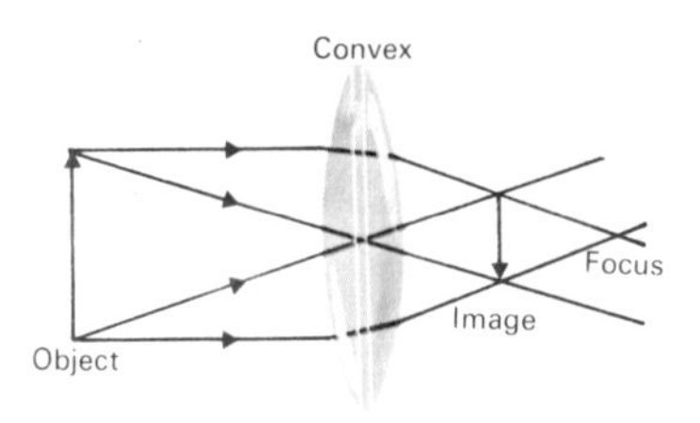

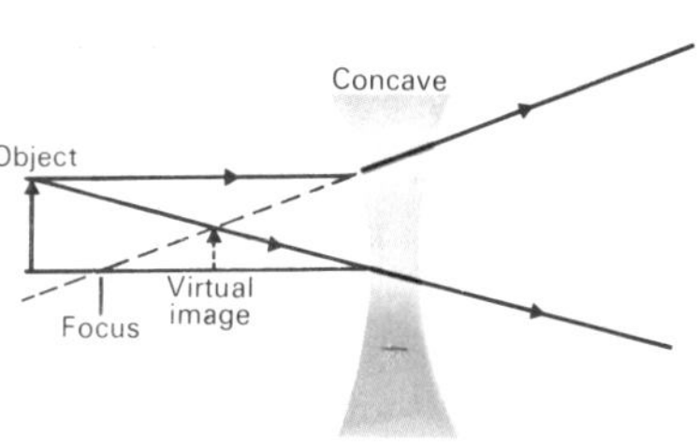

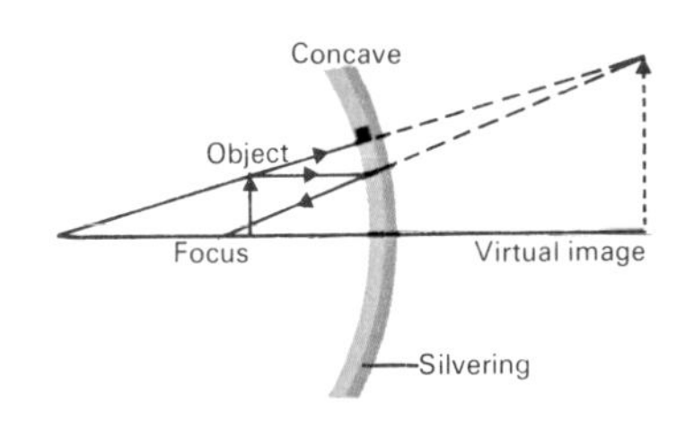

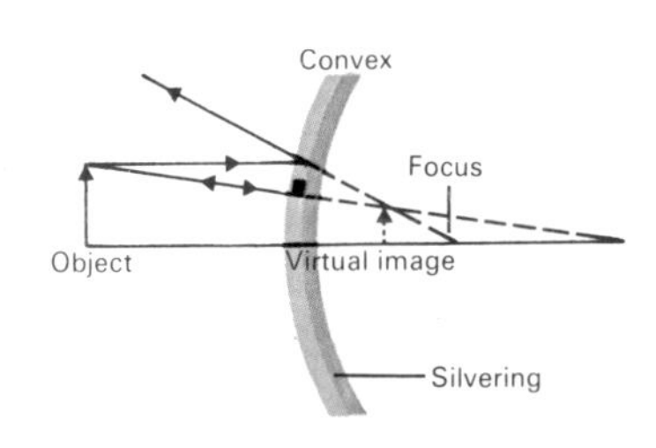

Structure of the human eye with schematic representation of short and long sightedness (*left*).

Image formation by lenses and mirrors (*right*).

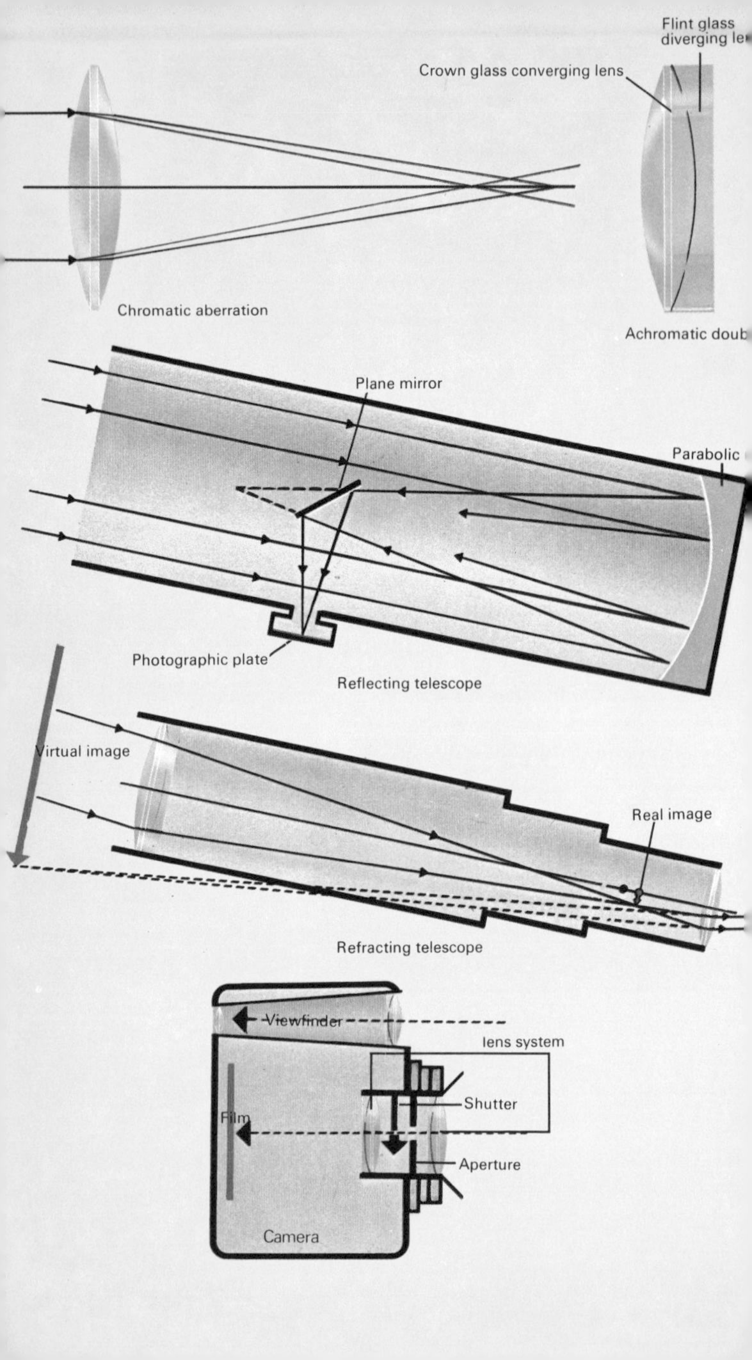
Flint glass
diverging le
Crown glass converging lens
Chromatic aberration
Achromatic doub
Plane mirror
Parabolic
Photographic plate
Reflecting telescope
Virtual image
Real image
Refracting telescope
Viewfinder
lens system
Shutter
Film
Aperture
Camera

Optical instruments

A camera consists essentially of a lens system which focuses light onto a light-sensitive film in a light-tight compartment. Light falls on the film only when the shutter is opened during an exposure. The amount of incident light is controlled by the size of the *aperture*, whose diameter is usually variable, and by the time of exposure (usually a fraction of a second). A suitable combination of aperture and exposure time can be selected for different light conditions by an exposure meter.

A good lens system consists of two *achromatic doublets* of fixed focal length, placed symmetrically about the aperture. The red component of light, refracted by a single convex lens, is focused further from the lens than blue light. This is a serious defect of lenses called *chromatic aberration*. It is corrected by cementing together two glass lenses of different refractive indices, one a converging lens the other weakly diverging. This achromatic doublet converges two different wavelengths to the same focus, thus reducing coloured fringing.

Most modern *astronomical telescopes* use a *reflecting system* in which a mirror collects the light from stellar bodies that are often so faint and tiny as to be invisible to the naked eye. Mirrors are free from chromatic aberration and can be accurately ground. A concave mirror of wide aperture focuses the light onto a photographic plate that can be exposed for several hours. An additional mirror system can be used to position the image more conveniently. The diameter of the mirror is made as large as possible to increase the intensity of the final image.

A simple refracting telescope consists of two convex lenses. Incident light (assumed parallel) is focused by the *objective* at its focal point. The *eyepiece* produces the final inverted virtual image. The magnification obtained is the ratio of the focal length of the objective to that of the eyepiece.

Prism binoculars consist of two refracting telescopes, one for each eye. Two sets of prisms, placed between each objective and eyepiece, invert the image by *total internal reflection* so that it is the right way up. An inverted image is unimportant in astronomy but for terrestial use the image must appear the right way up.

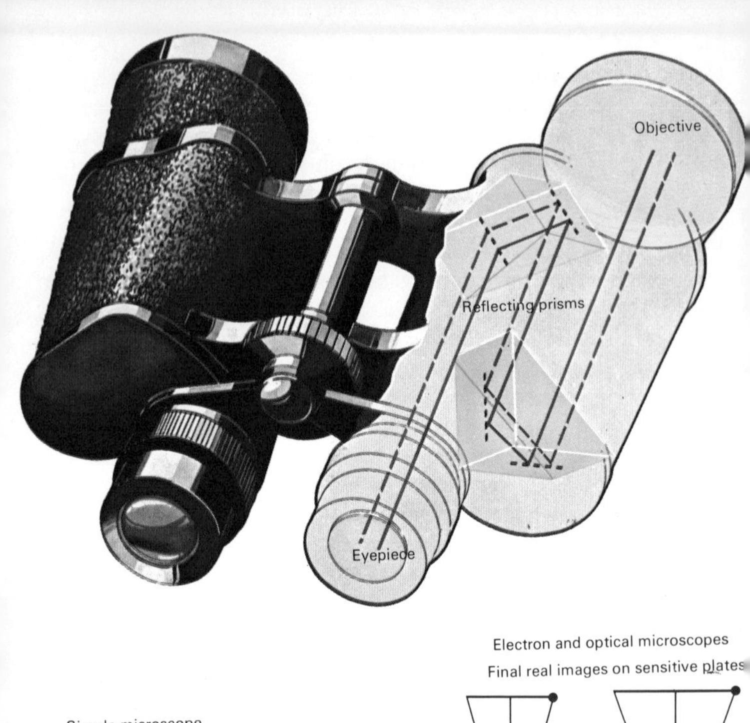
Objective
Reflecting prisms
Eyepiece

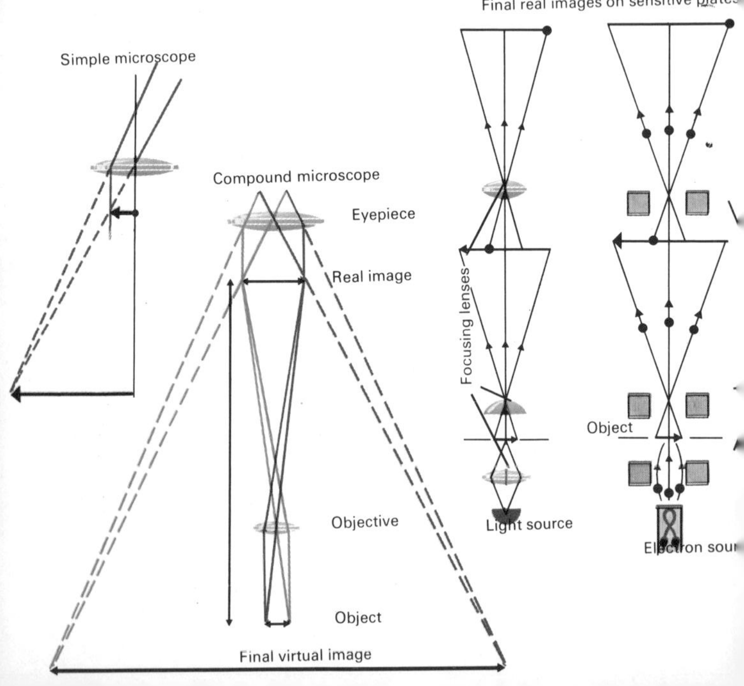
Electron and optical microscopes
Final real images on sensitive plates
Simple microscope
Compound microscope
Eyepiece
Real image
Focusing lenses
Object
Objective
Object
Final virtual image
Light source
Electron sou

The *microscope* provides an enlarged image of a small object. The *magnifying glass* is a convex lens of short focal length. The object is placed between focus and lens, and a virtual image is formed beyond the focus on the same side of the lens as the object. The magnification obtained is the ratio of the image size to object size.

Greater magnification is possible with the *compound microscope*. The objective forms an image at a point within the focal length of the eyepiece. The eyepiece consequently acts as a magnifying glass in producing the final image. The magnification is the magnification produced by the objective times that of the eyepiece. In practice, the object is placed on a thin, glass slide and illuminated from below by a light beam of high intensity. Both lenses are corrected for chromatic aberration and other lens defects. The maximum magnification, ranging from $\times$ 1000 to $\times$ 3000, is limited by the *resolving power* of the lens – the distance between two points that the microscope can just distinguish. As this distance is proportional to wavelength, a limit is set by an inherent property of light and no technical advance can improve it. The resolving power of the optical microscope is about 200 nm.

Using radiation of much smaller wavelength produces an enormous increase in resolution. The *electron microscope* works in a way analogous to the optical microscope except that a beam of electrons is used rather than light. A wave motion is associated with a moving electron so that electrons have similar properties to light waves, and as they can be deflected into a curved path by a magnetic field, an electron beam can be focused by such a field – called a *magnetic lens*. The wavelength associated with electrons depends on the voltage used to accelerate them. Voltages from 30 – 100 kV produce a wavelength of about 2 – 4 nm, a magnification of approximately $\times$ 1 000 000, and a resolving power less than one nm.

Prism binoculars
Simple and compound microscopes
Comparison of light and electron microscopes

Formation of spectrum from white light

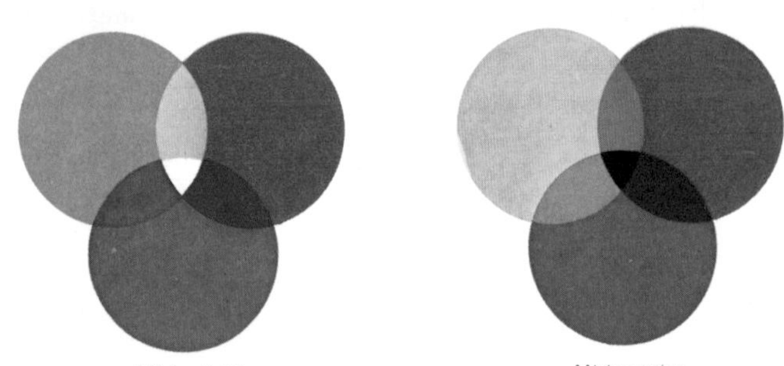

Mixing light

Mixing paint

(*Upper*) White light can be split into its component colours by a prism.

(*Lower*) The results of additive mixing of light (*left*) and subtractive mixing of pigments (*right*).

Colour

The velocity of light in a material medium (rather than in a free space) is different for different wavelengths. The refractive index therefore varies with wavelength, increasing in value as the wavelength decreases. Consequently, when a narrow beam of light enters a medium, such as a glass prism, it is split up into its component colours, red light being refracted less than violet light.

Not only can light be split up into its component colours, but these colours can also be combined, in the correct proportions, to produce white light. Light of only three colours is sufficient to synthesize white light. Red, green, and blue light give the best synthesis. Any colour can be obtained by a suitable combination of these three colours. This is the basis of both *colour photography* and *colour television*. Yellow light, for example, is produced by mixing green and red light.

When light enters a substance not all wavelengths are absorbed to the same extent. A part of the spectrum of the light falling on a coloured object is reflected; the remainder is selectively absorbed. The wavelength reflected by the object determines the colour as seen by the eye. White objects reflect all wavelengths, whereas black substances absorb almost all. A red carpet, illuminated by green light, appears black as green will be absorbed and no light will be reflected.

Mixing colour pigments, such as paints or dyes, is different from combining coloured lights. The final colour of a mixture of lights is the *sum* of the individual colours. The pigment in a paint absorbs part of the spectrum, reflecting only its particular colour. A mixture of paints absorbs (or subtracts) the wavelengths appropriate to its makeup. Combinations of red, yellow, blue, and white paint can produce any desired colour.

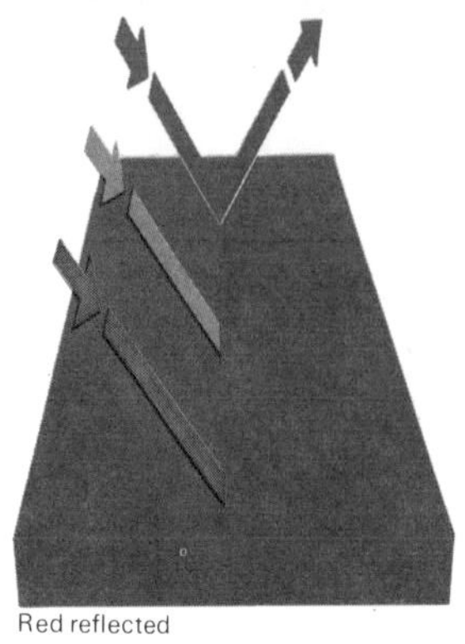

Red reflected
Green absorbed
Blue absorbed

Green reflected
Blue absorbed
Red absorbed

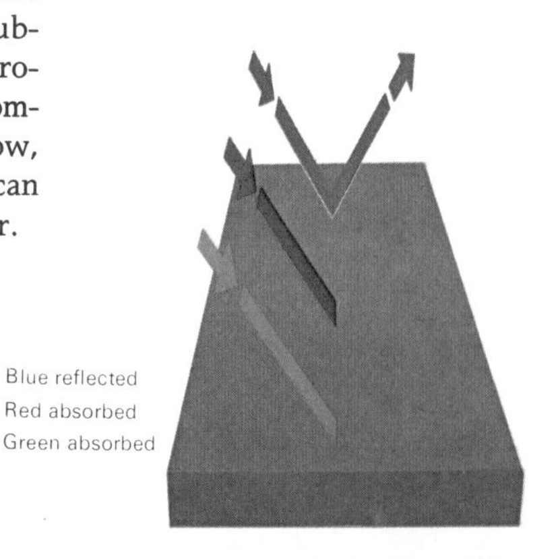

Blue reflected
Red absorbed
Green absorbed

Colours of opaque objects are produced by selective reflection.

Colour vision

The light-sensitive cells of the eye are of two kinds, *rods* and *cones*. The rods can respond to visible wavelengths of low intensity but cannot distinguish between different colours, therefore we have black and white vision at night. The cones respond to the higher intensities of daylight and as this response varies with wavelength, individual colours can be recognized. Colour vision can be explained by assuming that there are three separate systems of cones, sensitive to either red, green, or blue light. Incident light will therefore stimulate one or more of these systems to an extent that depends on its colour. Colour blindness, which is hereditary, results when one or more sets of cones are not functioning properly.

Colour television

In colour television, light is split into the three colours, red, green, and blue, by the use of special mirrors and filters. (A filter is a thin coloured film that is transparent only to light of its own colour.) Each colour is fed into a separate television camera tube so that the red, green, and blue content of the scene being televised is correctly converted into three separate signals.

These signals, following transmission, are received by three electron guns and the resulting electron beams are made to converge on a plate containing a large number of holes, the *shadow mask*. Behind the shadow mask lies the screen on which a very large number of phosphor dots are mounted in groups of three. One member of each trio will emit red light when activated by electrons, the second blue light, and the third green light, the intensity of the light emitted depending on the individual beam currents. As the three electron beams scan the mask they diverge through each of the holes; with correct focusing the one carrying the red signal strikes the red-emitting phosphor dot and likewise for the other two colours. Each cluster of dots appears as a single light source whose colour depends on the proportions of light emitted by the three phosphors. As the current of each electron beam varies, the colour changes accordingly and the original colours of the televised scene are reproduced.

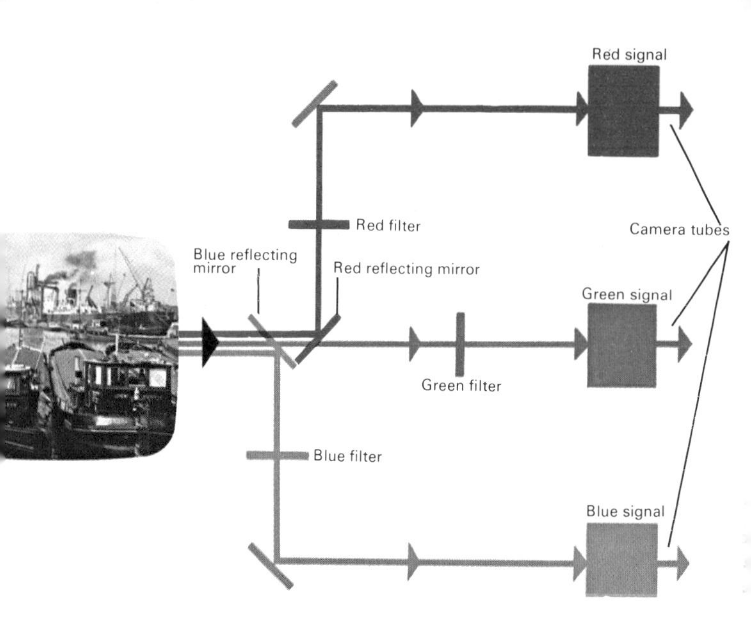

A colour television camera separates light into three components (*upper*). Transmitted signals control three electron beams which pass through a shadow mask to strike the appropriate phosphor dots.

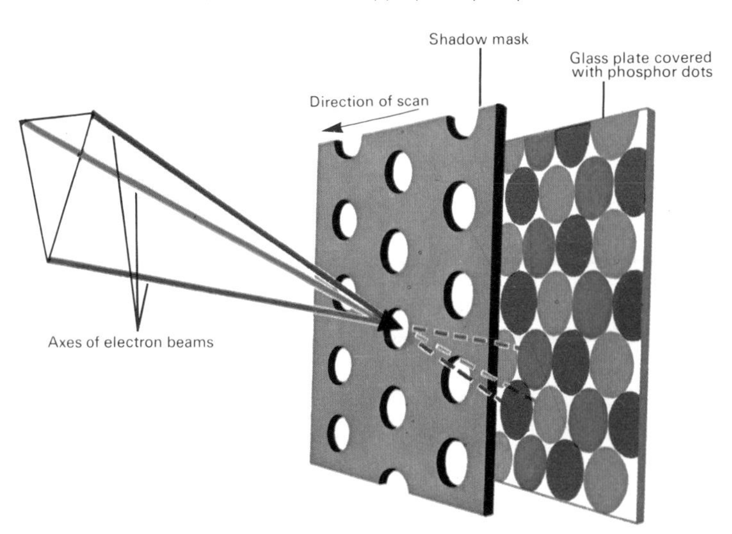

Wave theory of light

In 1680, Christiaan Huygens suggested that light travelled in the form of waves, but no definite evidence supporting this suggestion was found for over a century. Reflection and refraction could be satisfactorily explained in terms of rectilinear propagation, and it was not until *interference* effects were observed in 1801 by Thomas Young, that a wave theory became essential.

If two waves from the same light source, and with the same frequency, both arrive at a particular point after having travelled along different paths, they interfere with each other. The amplitude resulting from two overlapping waves at this point is the sum of the individual amplitudes. Two crests or two troughs will reinforce each other and the amplitude is doubled – *constructive interference*. A crest and a trough will cancel each other out and produce zero amplitude – *destructive interference*. The resulting interference pattern consists of alternate dark and bright bands. A bright band occurs if the path difference, d, equals $n\lambda$, where n is 1, 2, 3, ... ; a dark band occurs if $d=\frac{1}{2}n\lambda$. The colours observed in soap bubbles and in thin films of oil are interference effects.

Without wave theory, interference would be inexplicable. However, wave motion could not explain the interaction of

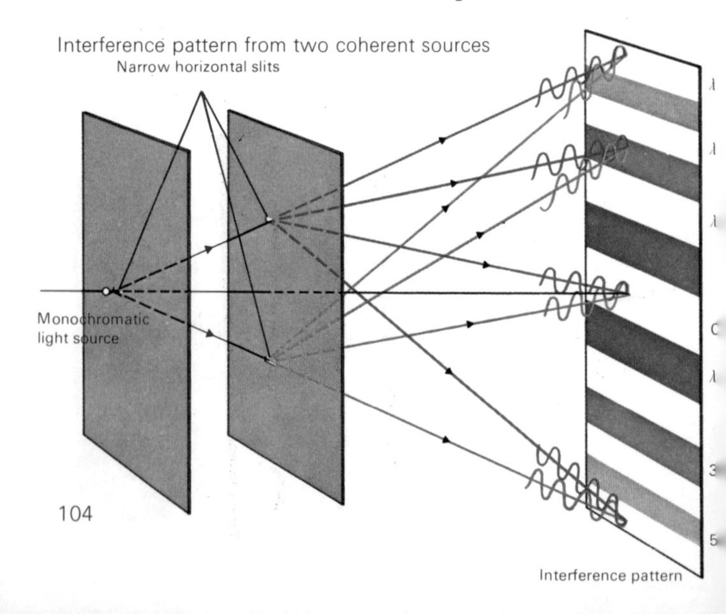

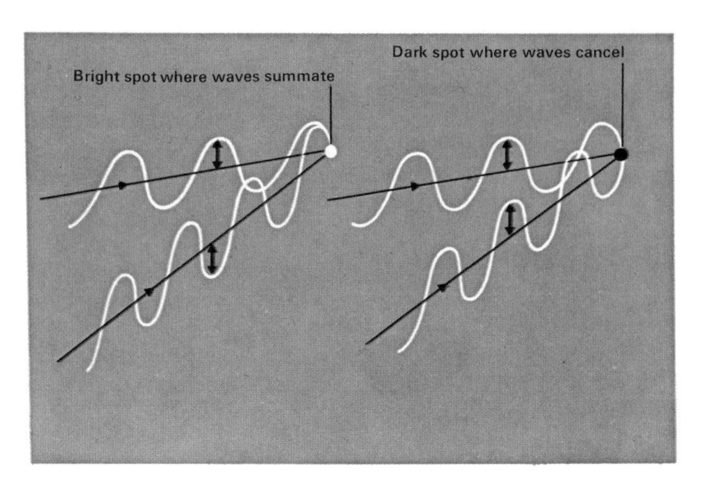

Formation of light and dark areas by constructive and destructive interference

radiation and matter. Planck postulated that energy exchange between radiation and matter involves an absorption or emission of a particle or *quantum* of energy. The *photoelectric effect* supported this idea, which led to a concept of radiation as a stream of quanta, called *photons,* all travelling at the speed of light. Depending on the circumstances, light can be considered either as a movement of waves or particles. This dual nature is not confined to radiation but also applies to particles, such as the electron. The wavelength, λ, associated with the motion of a particle is related to its momentum, p, (mass × velocity) by $\lambda = h\ p$, where h is the Planck constant. These ideas are basic to *quantum theory.*

The *Uncertainty Principle,* proposed by Werner Heisenberg, is one consequence of quantum theory. It states that it is impossible to perform accurate and simultaneous measurements of properties, such as the position, x, of a particle and its momentum, p; there is always an uncertainty in the values. The observation of a particle with a beam of light, for example, results in collisions between the photons and the particle. The particle is disturbed, its velocity altered, and hence its true momentum and position cannot be precisely determined.

The photoelectric effect

In the late nineteenth century, it was found that a clean plate of zinc acquired a positive charge when exposed to light or ultraviolet radiation. As the metal normally contains equal numbers of electrons and protons, the radiation must have caused an emission of electrons. This became known as the *photoelectric effect*. According to classical wave theory, the energy of a wave is independent of frequency, depending only on its intensity. The exchange of energy between the radiation and the electrons in the metal surface should therefore be greater as the intensity is increased. It was found, however, that increasing the intensity has no effect on the energy of the emitted electrons, but only produced a greater number of them. Their energy was only increased by raising the frequency of the radiation.

This apparently anomolous effect could only be explained by the quantum theory. By considering radiation as a stream of photons whose energy is proportional to the radiation

Photoelectric effect – electron energy depends upon frequency of incident light, whereas the number emitted depends upon light intensity.

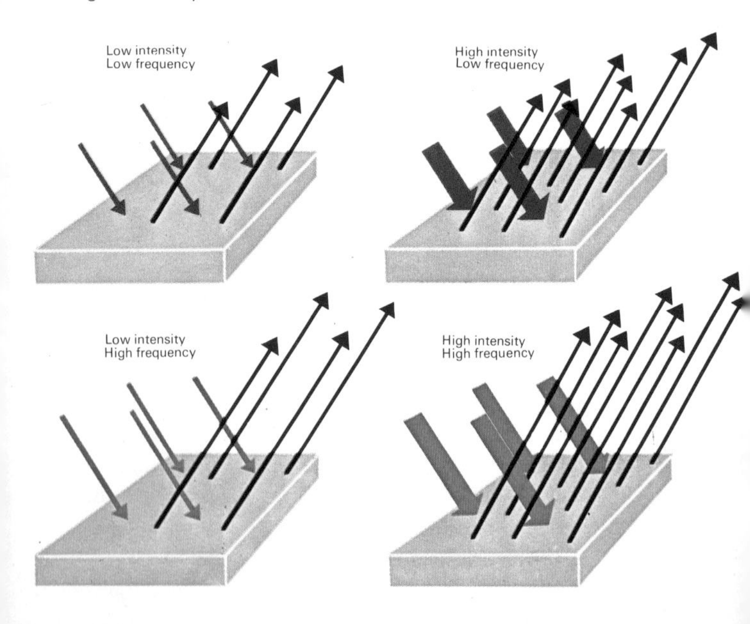

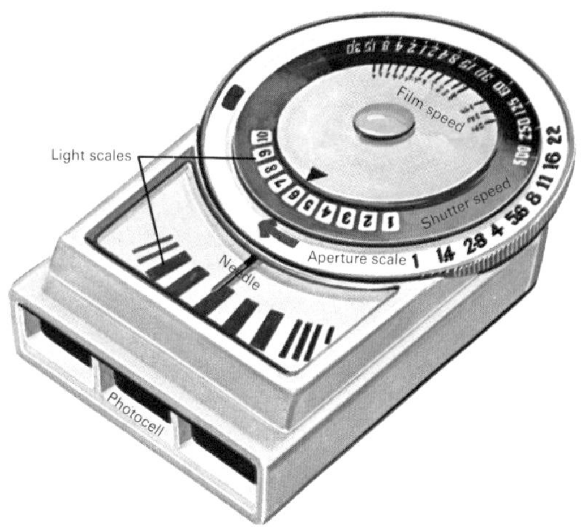

Exposure meter

frequency it follows that an increase in frequency would cause a greater exchange of energy between photons and electrons. The greater the intensity, the more photons would be available to release a larger number of electrons.

The photoelectric effect has practical applications. Once emitted, these photoelectrons can be attracted towards a positively charged anode and made to flow through an electric circuit. They therefore constitute an electric current whose value depends on the intensity of the radiation. A device using this principle, called a *photoelectric cell,* is used to measure the intensity of light or ultraviolet radiation. The *exposure meter,* used in photography, relies on the photoelectric effect of light incident on a surface of selenium or of cadmium sulphide.

Photons cannot be slowed down: in the photoelectric effect, they are completely annihilated. Photons differ from particles of matter, such as electrons, whose velocity can never reach that of light. Special relativity predicts that the mass, m, of a particle whose rest mass is m_0 increases as its velocity, v, increases: $m=m_0/\sqrt{(1-v^2/c^2)}$. The rest mass of a photon, travelling at the speed of light, c, is therefore zero, otherwise its mass when in motion would become infinite.

Emission spectra

The electrons in an atom are arranged in orbits, each orbit containing only a restricted number of electrons. As the number of electrons increases from one element to another, these orbits, which can be considered concentric with the nucleus, become filled up, the orbit nearest the nucleus being filled first. The energy of the atom is determined by which orbits are occupied by electrons, the energy increasing as the distance from the nucleus increases.

If an atom absorbs energy from its environment, an electron can jump into a higher energy orbit, if there is room for it. (This is only possible for the electrons in the outermost orbit, as the inner orbits contain their full quota). The atom is then in an *excited state*. In 1913, Niels Bohr postulated that radiant energy can only be absorbed or emitted by an atom in the form of photons, each photon having an energy *hf*. The absorption of one photon results in one electron jump. As an excited atom is unstable, this electron rapidly falls back to its

Quantum jumps from higher to lower orbits give rise to the different series of lines in the emission spectrum of hydrogen. (*Far right*) Comparison of continuous and line emission spectra

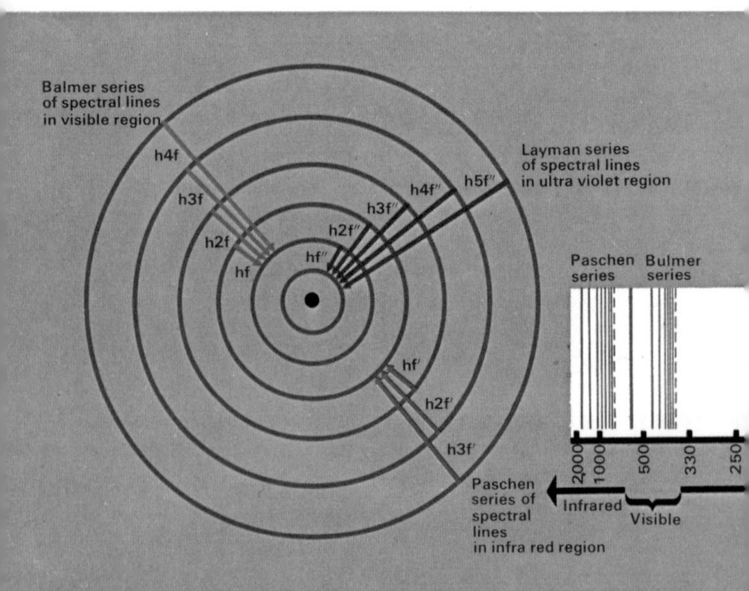

original orbit and the excess energy, equal to the difference between the high energy of the initial state and that of its final state, $E_i - E_f$, is released in the form of a photon. That is: $E_i - E_f = hf$. Due to the large number of possible orbits, many photons will be emitted, having a variety of frequencies. If the radiation emitted by a hydrogen atom falls on a photographic plate inside a *spectrograph* a series of sharp lines will be recorded, each line representing a different frequency. This is the *emission spectrum* of the atom. Each element has its own characteristic spectrum.

Gaseous elements usually have a line spectrum. In liquids there is a strong interatomic or intermolecular attraction, resulting in slight changes in the energy levels. The line structure is lost and bands appear. In solids the attraction is much greater, the width of the bands increases, and – if they overlap – a *continuous spectrum* is formed. The light emitted by a heated tungsten filament in an electric light is a continuous spectrum, covering the whole range of visible wavelengths. If an atom absorbs sufficient energy, an electron can escape from it altogether; the atom is then *ionized* and has a positive charge.

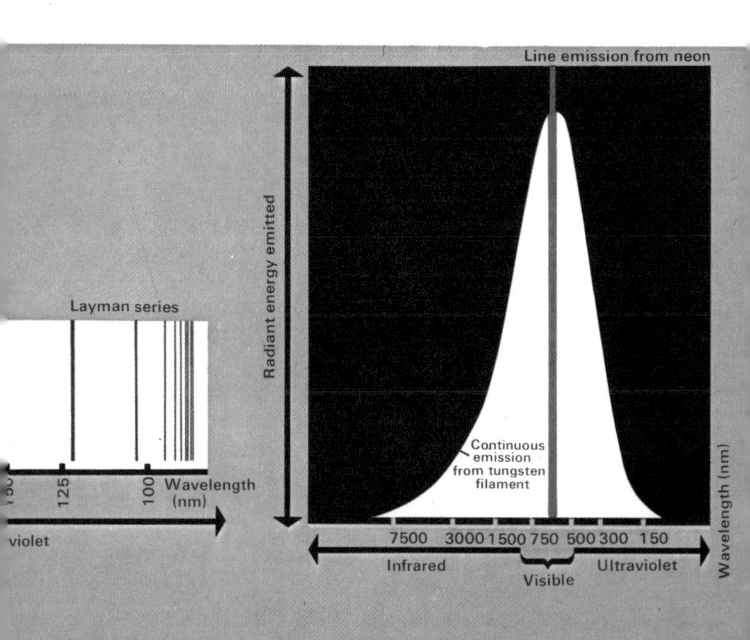

Absorption spectra

A substance that emits light of a certain frequency can also absorb light of that frequency. If a substance lies in the path of light or ultraviolet radiation, its atoms absorb those photons that correspond to the quantum jumps permissable to its electrons. The intensity of the light absorbed depends on the nature and thickness of the substance, and on the radiation frequency. All matter absorbs in some part of the visible or ultraviolet spectrum.

A *spectrograph* records both emission and absorption spectra. Radiation enters through a narrow slit and falls on to a quartz prism that splits the beam into its different frequency components. These are focused onto a photographic plate. If the original source of light has a continuous spectrum of constant intensity the absorption spectrum, formed by passing the radiation through a gaseous medium, consists of dark lines corresponding to the frequencies absorbed by the

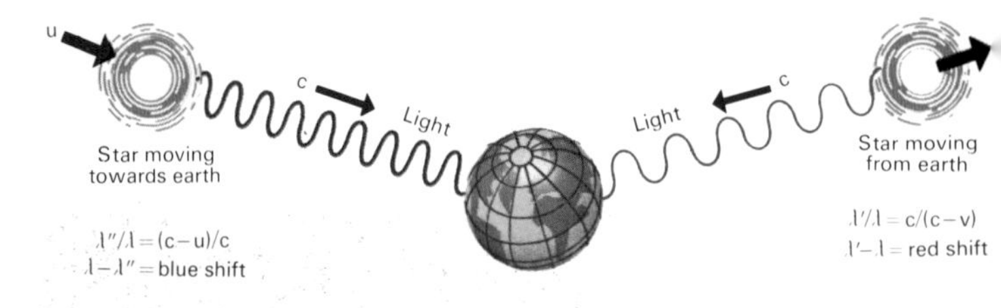

(*Above*) The apparent wavelength of light depends upon the relative motion of source and observer.
(*Opposite above*) Fraunhofer lines in solar spectrum
(*Opposite below*) Comparison of absorption and emission spectra

atoms of the medium. These lines can merge into bands for the same reasons as in emission spectra.

In 1814, Joseph von Fraunhofer observed that the sun's continuous spectrum was crossed by hundreds of dark lines, known as *Fraunhofer lines*, indicating that certain elements, including hydrogen and helium, are present in the sun's atmosphere. The centre of the sun provides the continuous spectrum and its surrounding atmosphere acts as the absorbing medium. Every star has its own characteristic absorption spectrum that gives us information about the star's velocity away from or towards the earth. Due to the Doppler effect, the whole spectrum from a moving star is shifted towards the red or the blue end of the visible spectrum.

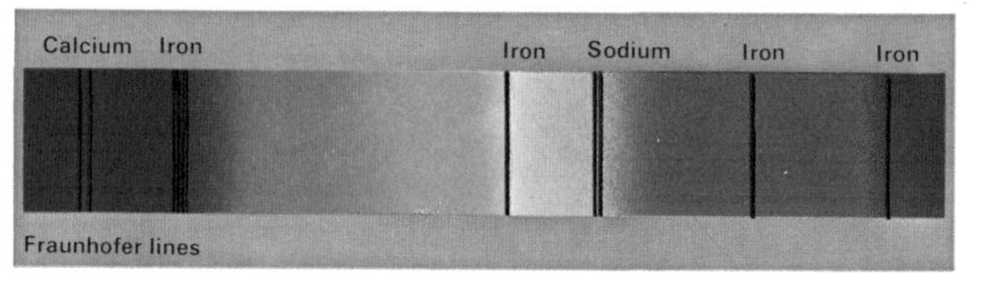

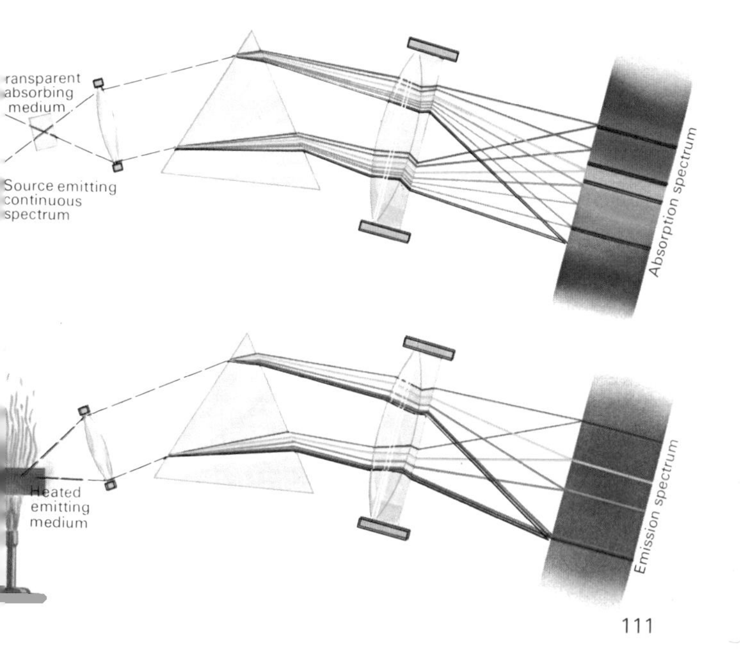

The laser

The emission of a photon by an excited atom or molecule, cannot normally be controlled and may contain photons of many different frequencies, emitted in all directions. In 1917, Einstein predicted that emission could be induced or *stimulated,* by the interaction of an incoming photon of suitable energy. For *stimulated emission* to be of use, there must be a large number of electrons in a particular high energy level. This condition, called *population inversion,* is achieved by pumping energy into the system. If an atom has two high-energy states, electrons in the highest state, E_n, are stimulated to jump to a lower state, E_m, by the action of photons of low intensity electromagnetic radiation having an energy $(E_n - E_m)$. Photons resulting from the emission process travel in approximately the same direction as the stimulating radiation and have the same frequency, $(E_n - E_m)/h$. The emitted photons can themselves also stimulate emission so that an amplified and narrow beam of monochromatic (single frequency) radiation is produced, which has a high energy/unit area. The greatest advantage of stimulated emission however is that it is coherent (in phase); spontaneous emission is incoherent.

Ruby laser

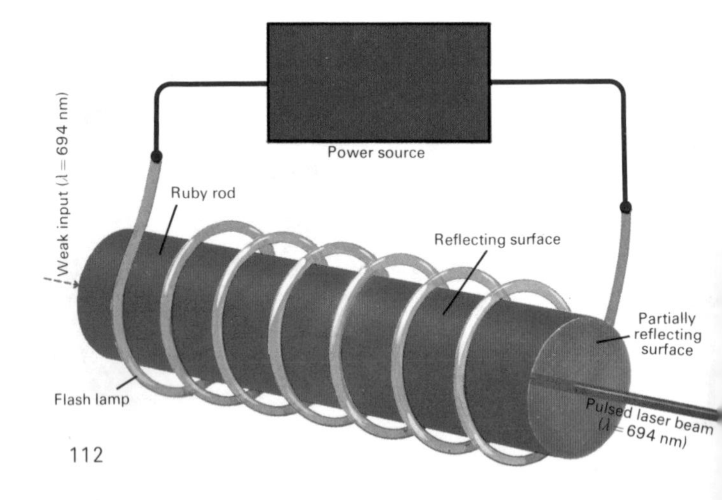

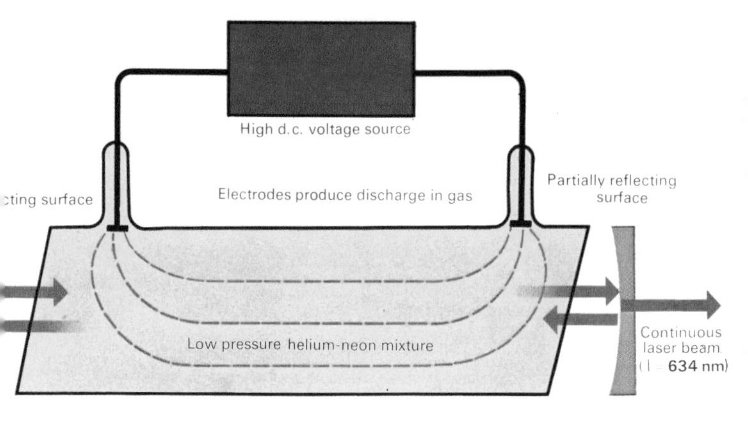

Helium-neon gas laser

Stimulated emission is used in the *laser* to amplify light, ultraviolet and infrared radiation. The lasers so far developed can be divided into three categories: solid-state, gas, and semiconducting lasers. In the solid-state ruby laser, chromium ions (present as an impurity in aluminium oxide) are stimulated to emit photons, having been previously excited by an extremely intense flash of light. The stimulated radiation, which is pulsed, has a wavelength of 694.3 nm. In a gas laser the low-pressure gas is excited by a continuous electrical discharge passing through it. The helium-neon laser produces a continuous visible beam at 634.2 nm; the carbon dioxide laser has a more powerful beam in the IR at 10.6 μm.

The laser radiation is made to oscillate between two mirrors – one reflecting, the other semi-reflecting – at either end of the laser. The distance between the mirrors is $\frac{1}{2}n\lambda$, where n is an integer and λ is the laser beam's wavelength. Only this wavelength is made to oscillate or resonate. The system is called a *resonant cavity*. The beam is greatly amplified by additional stimulated emission during its many reflections. The beam emerging through the semi-reflecting mirror is therefore very powerful.

ULTRAVIOLET RADIATION

The human eye does not respond to wavelengths less than that of the extreme violet. The ultraviolet (UV) region of the electromagnetic spectrum lies beyond the visible region, having a wavelength range from 380 nm to approximately 10 nm. About 5 per cent of the radiation emitted by the sun is UV. Although this is not a large proportion, the energy of this radiation is important because of its high frequency. The energy exchanged between UV and the atoms and molecules of matter causes ionization – the liberation of electrons, leaving positively charged ions. UV is only one type of *ionizing radiation*: *X-rays* and *γ-rays* have much greater ionizing powers.

Ionization of molecules, such as oxygen, in the ionosphere is largely caused by UV from the sun and results in the formation of ozone (O_3). This ozone layer in the upper atmosphere (together with dust in the lower atmosphere) absorbs a considerable amount of UV, so that only a small fraction penetrates to ground level.

The higher wavelengths of UV, down to 200 nm are not absorbed and can penetrate the atmosphere. These wavelengths have a beneficial effect on health. Certain reactions can only take place in the presence of UV, such as the forma-

Fluorescent lamp

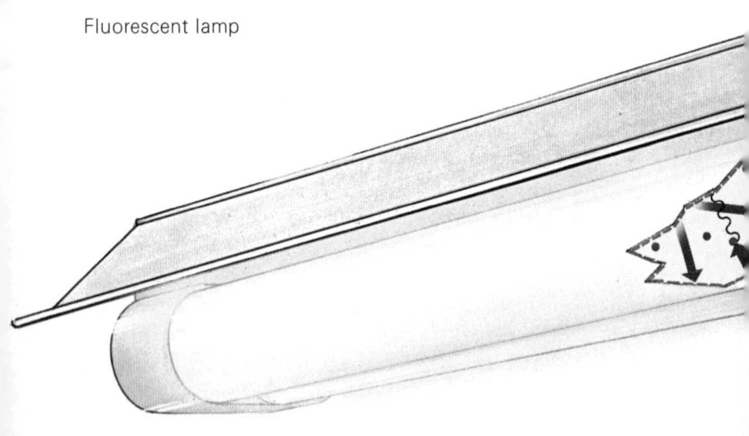

tion of vitamin D. A narrow band of wavelengths around 300 nm stimulates the production of melatonin in the skin producing a sun tan.

X-rays

The X-ray section of the electromagnetic spectrum lies between the UV and γ-ray regions, having an approximate wavelength range of 10 to 0.1 nm. X-rays were discovered in 1895 by Wilhelm Roentgen. They are produced when high speed electrons strike the atoms of the heavier elements. The resulting exchange of energy between the incident electrons and the electrons in the inner shells of the target atoms causes the electrons in the atom to jump from one orbit to another. X-rays are emitted when the electrons return to their original state.

A modern X-ray tube is evacuated and contains a tungsten filament and a tungsten target. The filament is heated to incandescence by an electric current, liberating a stream of electrons. These are accelerated towards the target by means of a high potential difference maintained between the cathode

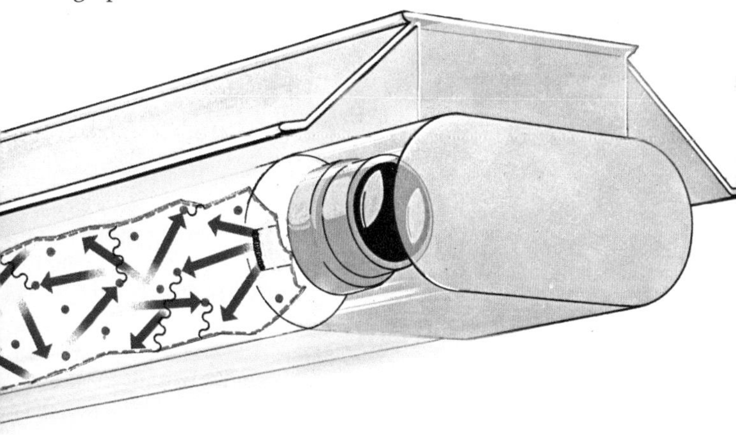

Phosphor coating — Atoms of mercury vapour — Electrons emitted from filament — Ultraviolet radiation emitted by mercury atoms

and the target anode. This voltage, which can vary from 1 kV to 1 MV, controls the energy of the X-rays. The energy of the accelerated electrons equals the voltage, V, times the electron charge, e. If all this energy is given to the target atoms, the frequency of the resulting X-rays is eV/h. The X-ray intensity is controlled by the filament current and the target material.

One phenomenon associated with the wave motion of both X-rays and light is *diffraction*. If radiation is passed through a narrow slit, the edges of the slit act as two coherent sources (same frequency and phase) and the different distances

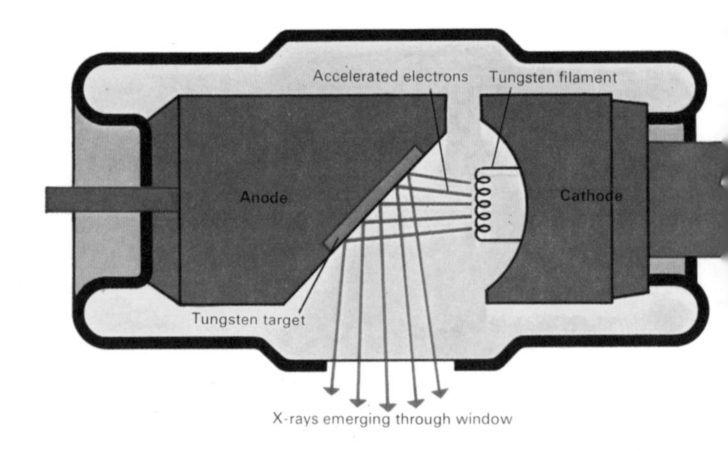

X-ray tube

travelled by waves from these points produce a diffraction pattern where they converge, in the same way as an interference pattern is produced. The slit must be much smaller than the radiation wavelength. Early in the twentieth century it was found that the interatomic distances of approximately 0.1 nm in crystals are of the right order to produce X-ray diffraction effects.

If X-rays fall at an angle, k, onto a series of parallel atomic planes a distance, d, apart, the beam will be scattered in all directions by each plane. This scattered beam will have its maximum intensity at an angle, k. If the path difference, $2d \sin k$, between two waves scattered from two adjacent planes, is an even number of wavelengths, a bright spot on the diffraction pattern is produced; an odd number of wavelengths produces a dark spot. X-ray diffraction is used to determine the atomic structure of crystals by analysis of these patterns. The study of *X-ray crystallography* has developed rapidly and it is now possible to determine the crystal structures of extremely complicated biochemical molecules such as haemoglobin and insulin.

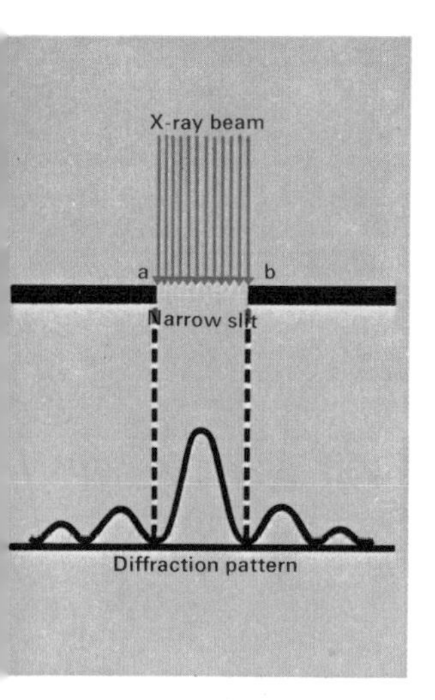

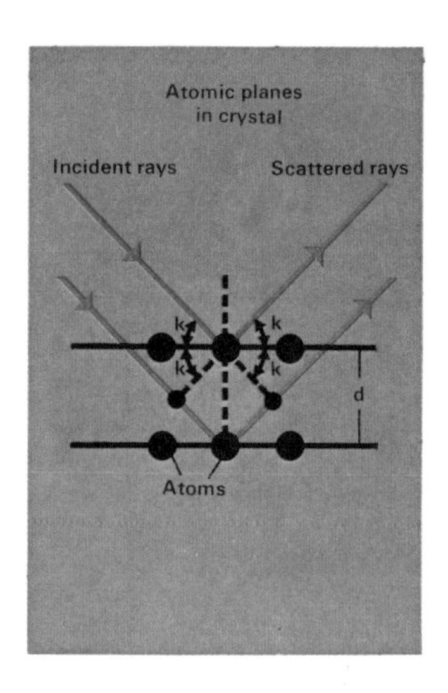

Production of diffraction pattern by X-ray beam
X-ray diffraction in a crystal

Medical uses of X-rays

X-rays have several medical uses. Body tissue is largely composed of relatively light atoms – hydrogen, oxygen, carbon, and nitrogen. These atoms do not absorb X-rays to the same extent as heavier atoms, such as calcium which occurs in bones. An X-ray photograph (*radiograph*) is obtained when part of the body is placed between a low-energy X-ray source

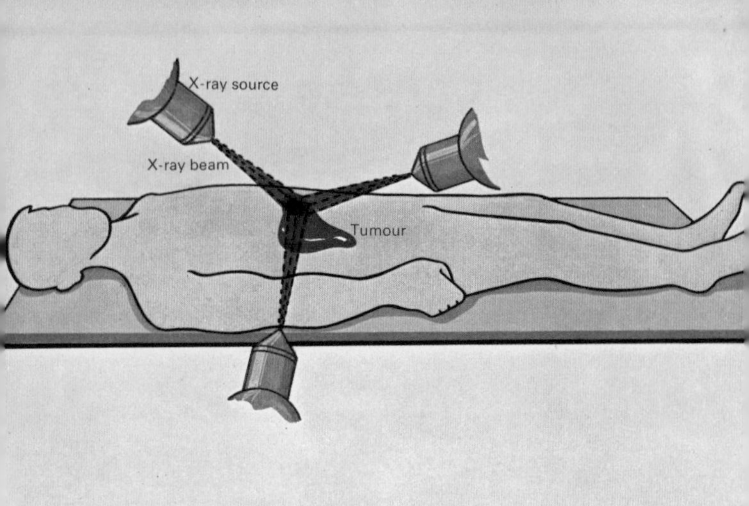

(*Above*) Irradiation of tumour – maximum radiation is received only by affected tissue.

(*Below*) Cosmic rays produce elementary particles in the upper atmosphere.

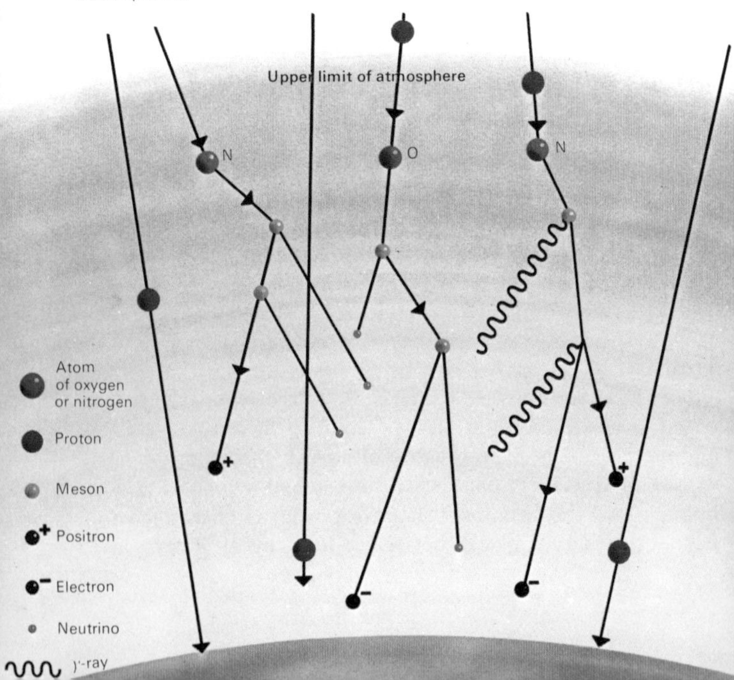

and a photographic plate. Due to the high X-ray absorption, bone structure can be clearly seen on the radiograph and a broken bone or bone defect can be diagnosed.

Higher-energy X-rays (or electrons) are used to treat cancer. A highly overactive cell – a cancer cell – is far more susceptible to damage or death by irradiation than a cell that is reproducing at a normal rate. Once diagnosed, a tumour (a concentration of cancer cells) is exposed to a carefully-directed narrow beam of X-rays; the tumour can be irradiated from two or three directions so that damage to intervening tissue is minimized and the maximum concentration of X-rays is received by the tumour.

About 75 per cent of the body is composed of water and many of the effects of radiation can be explained by its ability to ionize water molecules in its path. The ionization products of water are extremely reactive and will attack the vital molecules that control the functions of a cell. The resulting damage prevents these molecules from working properly and the cell will die. Radiation damage also occurs when the cell molecules themselves are ionized, again interfering with their function. An indiscriminate dose of ionizing radiation can however, produce cancer by excess damage to the body cells. A very low dose of radiation might not have a disastrous effect on a person, but it can damage the radiation-sensitive germ cells (eggs or sperms) of the body, and this damage will be passed down to another generation resulting in a range of ill-effects. The use of X-rays must be strictly controlled.

Cosmic radiation

Extremely high-energy radiation reaches the earth from outer space. This is called *cosmic radiation* and consists principally of positively-charged particles – mainly *protons* but also of the nuclei of helium atoms (α – particles) and heavier nuclei, such as that of iron. After entering the earth's atmosphere the high-energy protons collide with oxygen or nitrogen atoms. These collisions produce other elementary particles, which are all unstable. One such particle is the *meson* of which there are several types. These all eventually decay to *γ-rays, electrons, positrons* and *neutrinos*. Positrons have the same mass as an electron, but have a positive charge.

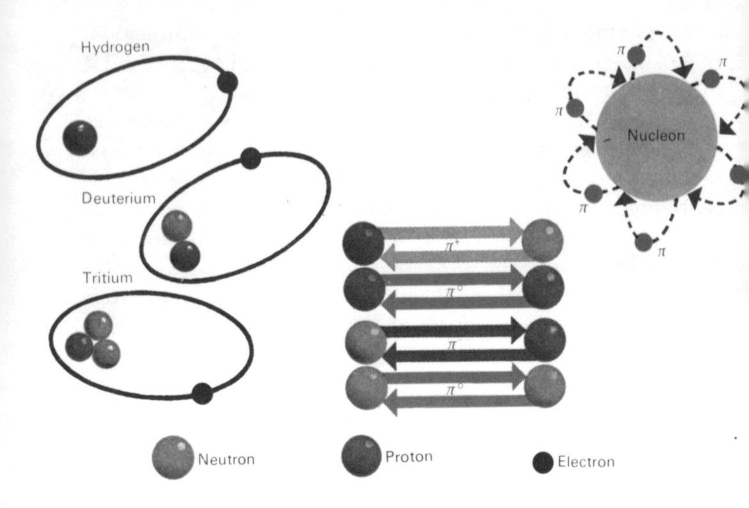

The isotopes of hydrogen (*left*). Unstable π-mesons are thought to be exchanged between nucleons (*centre*), thus serving to bind the particles of the atomic nucleus together (*right*).

Table of elementary particles

Group	Name	Symbol	Rest mass	Spin	Mode of decay	Half-life (secs)
	Photon	γ	0	1	Stable	Stable
Leptons	Neutrino	ν	0	$\frac{1}{2}$	Stable	Stable
	Electron	e^-	1	$\frac{1}{2}$	Stable	Stable
	Positron	e^+	1	$\frac{1}{2}$	Stable	Stable
	μ-meson	μ^+ μ^-	207	$\frac{1}{2}$	$\mu^- \rightarrow e^- + 2\nu$ $\mu^+ \rightarrow e^+ + 2\nu$	$1{\cdot}52 \times 10^{-6}$
Mesons	π-meson	π^+	273	0	$\pi^+ \rightarrow \mu^+ + \nu$	$1{\cdot}8 \times 10^{-8}$
		π^-	273	0	$\pi^- \rightarrow \mu^- + \nu$	$1{\cdot}8 \times 10^{-8}$
		π°	264	0	$\pi^\circ \rightarrow 2\gamma$	$6{\cdot}16 \times 10^{-17}$
Nucleons (Baryons)	Proton	p	1836	$\frac{1}{2}$	Stable	Stable
	Neutron	n	1838	$\frac{1}{2}$	$n \rightarrow p + e^- + \nu$	700

NUCLEAR ENERGY

Until 1911, it was thought that the atom consisted of a homogeneous mixture of electrons and protons. Ernest Rutherford then proposed that the positively charged protons are concentrated at the centre of the atom inside a tiny core called the *nucleus* and that the negatively charged electrons orbit around the nucleus. The electrons and protons have an equal but opposite charge. As an atom is normally neutral the number of electrons and protons must be equal; this number is the *atomic number, Z*. The proton is 1836 times heavier than the electron, so that the atomic mass is concentrated in the nucleus, which occupies a minute volume compared to that of the whole atom.

All atomic nuclei with the exception of the hydrogen nucleus, contain another particle called the *neutron*. This particle was discovered in 1932 and it was found to have a mass very close to that of the proton, namely 1838 times the electron mass. The neutrons and protons, known collectively as *nucleons*, are bound together very tightly inside the nucleus. Atoms with the same number of protons, but a different number of neutrons, are called *isotopes*. Hydrogen contains one proton only; the isotopes of hydrogen are *deuterium* (one proton and one neutron) and *tritium* (one proton and two neutrons). As the chemical properties of an element are determined by the number of electrons in the atom, and consequently by the number of protons in the nucleus, an increase in the number of neutrons in the nucleus has no effect on the element's chemical properties.

Bound inside the nucleus, the neutron is stable. Outside the nucleus, however, an unbound neutron loses its stability and disintegrates (or *decays*) in about 12 minutes, into a proton, an electron and another particle called a *neutrino*. This disintegration can also occur in an unstable radioactive atom. The neutrino has zero charge and zero rest mass.

The nucleons are very tightly packed in the nucleus, the radius of which varies from $2-9 \times 10^{-15}$ metre. The protons are repelled electrostatically by their similar charge, but this repulsion is overcome by a much greater force which only acts over a very small distance. This force, called an *exchange force,* binds the nucleons together very tightly. An elementary particle, the *π-meson,* is thought to be exchanged back and forth between nucleons in a very short time – less than 10^{-34} seconds. This exchange is considered responsible for the nuclear force. The π-mesons (or *pions*) have either a positive, negative, or zero charge, and each one is unstable. The *μ-meson* (or *muon*) is a decay product of the pion. Both pions and muons are detected in cosmic radiation.

The binding energy of nuclei

The mass of nuclear particles is extremely small but it can be determined with great accuracy. The *relative atomic mass,* (or *atomic weight*), *M,* is the mass of an atom with reference to an isotope of carbon containing twelve nucleons, carbon-12, the mass of which is defined as 12 atomic mass units (amu). As an element normally has several isotopes, and the atomic mass is the average of the mass of the isotopes in the proportions in which they occur naturally, the atomic mass is never a whole number. The *mass number, A,* is the whole number nearest to the atomic mass, and for a proton or neutron equals one. The mass number of an isotope of a particular element represents the number of nucleons in its nucleus. An isotope, *X,* is usually written ${}^{A}_{Z}X$, where *Z* is the atomic number – the number of protons – and $(A - Z)$ is the number of neutrons. Carbon-12 is written ${}^{12}_{6}C$, or simply ${}^{12}C$.

The atomic mass of the proton is 1.00782 amu and that of the neutron 1.00867 amu. However, the mass of a nucleus is not the sum of the masses of its constituents, but slightly less than this. A deuterium atom, containing one proton and one neutron, has an atomic mass of 2.01410 amu – a loss in mass of 0.00239. This is explained by Einstein's principle of the *equivalence of mass and energy.* This states that the mass, *m,* is equivalent to an amount of energy, *E,* the relation being: $E=mc^2$, where *c* is the velocity of light. The apparent loss of mass, observed in every nucleus except hydrogen, is equivalent to the energy released when the nucleons combine together to form a nucleus. It is consequently called the *binding energy* of the nucleus, E_B, and is equal to the sum of the masses of the unbound nucleons minus the atomic mass, *M*: $E_B=(Z\times \text{proton mass}+(A-Z)\times \text{neutron mass}) - M$. In

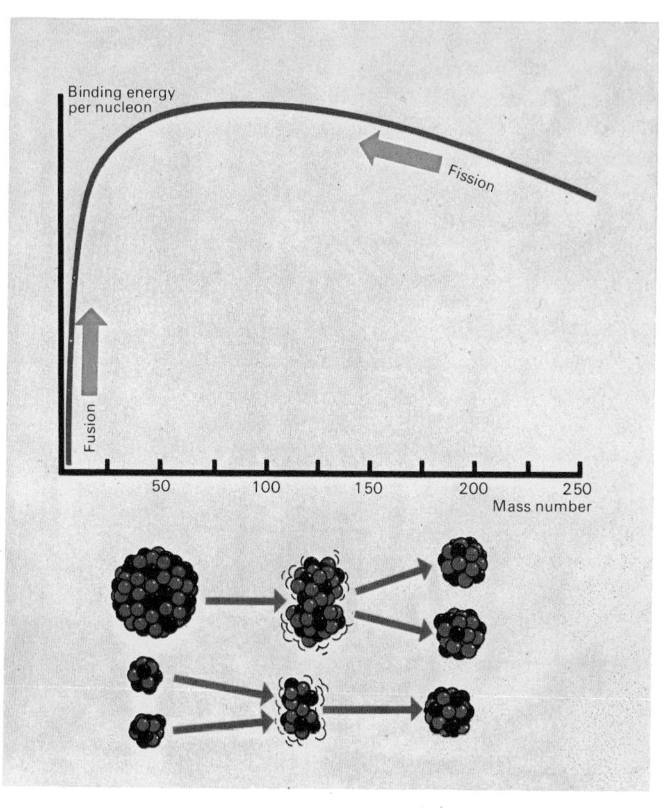

Binding energy per nucleon plotted against mass number

order to split a nucleus, it must be supplied with its binding energy (except in the case of spontaneous radioactive disintegration).

Radioactivity – stability and instability of atoms

The stability of a nucleus depends on how great its binding energy is. For light atoms, the most stable state is associated with equal numbers of protons and neutrons – as with helium (two protons, two neutrons). As the atomic mass increases, the repulsion between protons becomes more important and the more stable heavy atoms tend to have more neutrons than protons.

Some isotopes, known as *radioactive isotopes,* are unstable and emit radiation spontaneously. This radiation is of three types; one is a very penetrating (hard) radiation, named *beta-particles* (β-particles) another is easily absorbed by matter (soft) and is called *alpha-particles* (α-particles). There is often an additional emission of high-energy γ-rays accompanying one or other type of radiation.

A β-particle is an energetic electron, emitted at high speed from a nucleus. The loss of an electron results in an increase of one in the nuclear charge. The atomic number, Z, therefore increases by one, although the mass number, A, remains unaltered: for example $^{32}_{15}P$ (phosphorus) $\xrightarrow{\beta}$ $^{32}_{16}S$ (sulphur). As a nucleus does not contain electrons, a β-particle arises from the decay of a neutron into a proton, electron, and

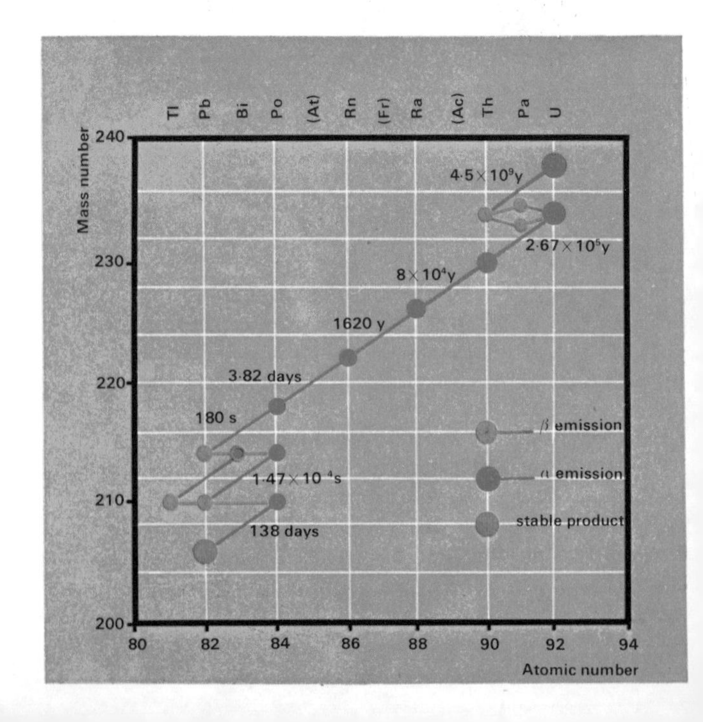

The series formed by the decay of ^{238}U

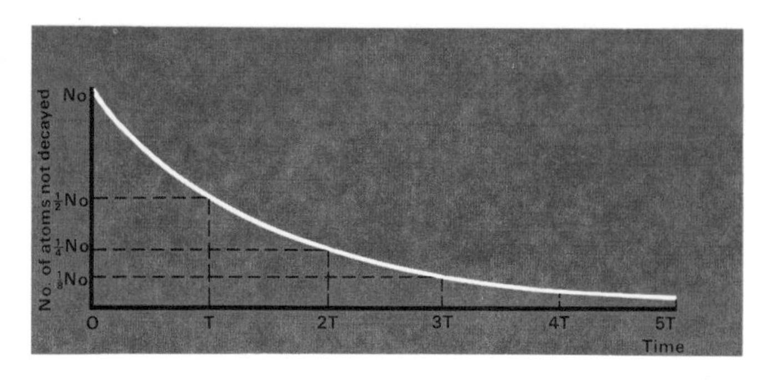

The decay curve of a radioactive isotope

neutrino. If β-particles are emitted alone their kinetic energy should be constant, being equivalent to the difference in mass of the parent isotope and its decay product. However, as a wide range of energies is found, it was postulated that a neutrino is emitted concurrently with the electron: the total energy of electron and neutrino is constant, although their individual energies can vary. The emission of the electron-neutrino pair from a nucleus is analogous to the emission of a photon from an excited atom. The neutron can be considered as a high-energy state of a nucleon and the proton as a lower energy state of the same nucleon.

α-particles are positively-charged helium nuclei. Following α-emission, a nucleus is transformed into one having four fewer nucleons and a charge decreased by two: for example: $^{226}_{88}Ra$ (radium) $\xrightarrow{\alpha}$ $^{222}_{86}Rn$ (radon).

Each radioactive isotope (radioisotope) decays at a different rate that depends on the number of atoms of the isotope present; it is independent of chemical or physical conditions. As the number of atoms that have not decayed decreases, the number of disintegrations decreases. The *half-life* of an isotope is the time taken for half the number of atoms to decay. After twice the half-life, half the remaining atoms will have decayed leaving one quarter of the original atoms. Half-lives can vary from less than a microsecond up to 10^{10} years.

Practically all naturally-occurring radioisotopes have

atomic numbers between 81 and 92, and can be grouped into three series – the *uranium, thorium,* and *actinium series*. Each member is a disintegration product of the preceding isotope in the series, resulting from α– or β–emission. The last member is not radioactive.

Artificial radioactivity

It is possible to produce artificial radioisotopes by bombarding certain atoms with high energy neutrons. A naturally-occuring element can be exposed to a stream of neutrons inside a nuclear reactor. Some of these neutrons are captured by the nuclei of the target element. Cobalt-60 (27 protons, 33 neutrons) is produced from cobalt-59 (27 protons, 32 neutrons) in this way. Because of the additional neutron, these isotopes are unstable and therefore radioactive.

Radioisotopes have a variety of uses. Many of them are used in the diagnosis and treatment of diseases such as cancer. A radioisotope has the same chemical properties as the stable isotopes of the same element and will therefore follow the same path through the body. Certain elements normally accumulate in particular parts or organs. If the cells in such an area are highly overactive – for example, cancer cells – there will be an increased accumulation of the relevant element. A radioisotope can indicate the presence of a tumour, as its high concentration will produce a much greater number of disintegrations than would be expected from normal tissue. The radioisotopes of iodine, ^{131}I and ^{125}I, like natural iodine, tend to accumulate in the thyroid gland (and also in the kidneys) and can register overactivity or underactivity in these organs. Radioisotopes used in diagnosis must have a relatively short half-life so that they decay rapidly, and must emit β– or γ-rays of sufficient energy to be detected by an external counter. Radiation emitted by high concentrations of isotopes can kill cancer cells, and can be used externally – on skin cancer, say – or internally.

A radioisotope, such as tritium, ^{3}H, can be introduced into a chemical substance (then said to be *labelled*), so that its path and concentration can be studied in a normally inaccessible environment. The formation and reactions of biochemicals inside the body can be studied in this way.

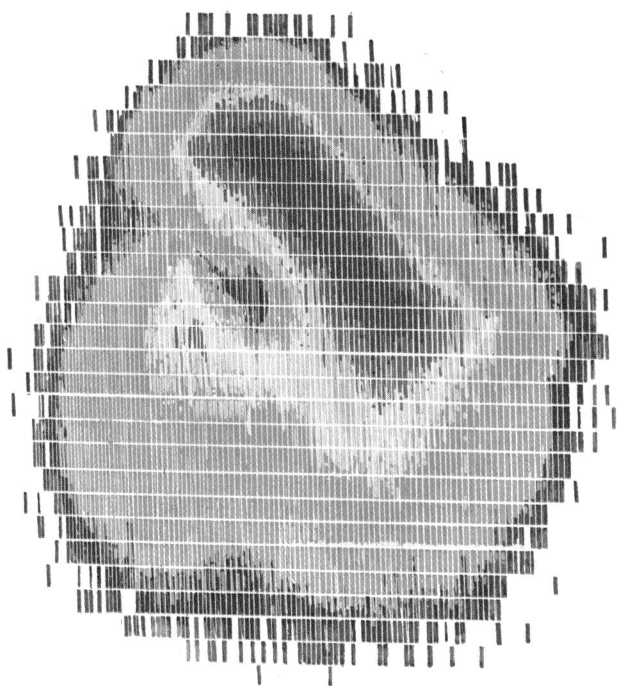

A scanner plots the concentration of radioactive isotope taken up by an overactive tissue.

A source of radioactivity can be used to control the thickness of a continuously produced sheet of metal.

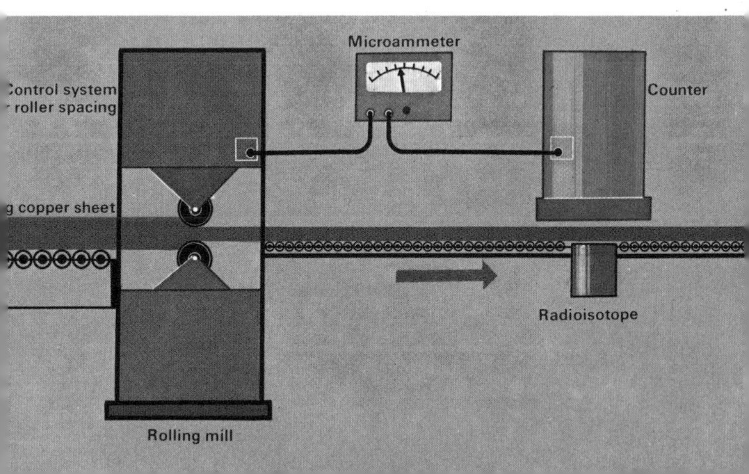

The transuranic elements and their production

The naturally-occurring elements occupy the first ninety-two places in the periodic table, uranium having the greatest atomic number of 92. However, elements having an atomic number greater than 92 can be artificially produced. These are the *transuranic elements*, created by bombarding heavy elements such as uranium, with high-energy particles – neutrons, protons, etc. – some of which are captured by the target elements. Thirteen transuranic elements have been made to date, with atomic numbers up to 105. All their isotopes are radioactive so that they possibly existed on earth millions of years ago but have since decayed. Most of them are produced in very small quantities and have not found much application, except for *plutonium* (atomic number 94) – a key material in nuclear weapons and reactors.

In a linear accelerator electrons are accelerated during successive halves of the voltage cycle.

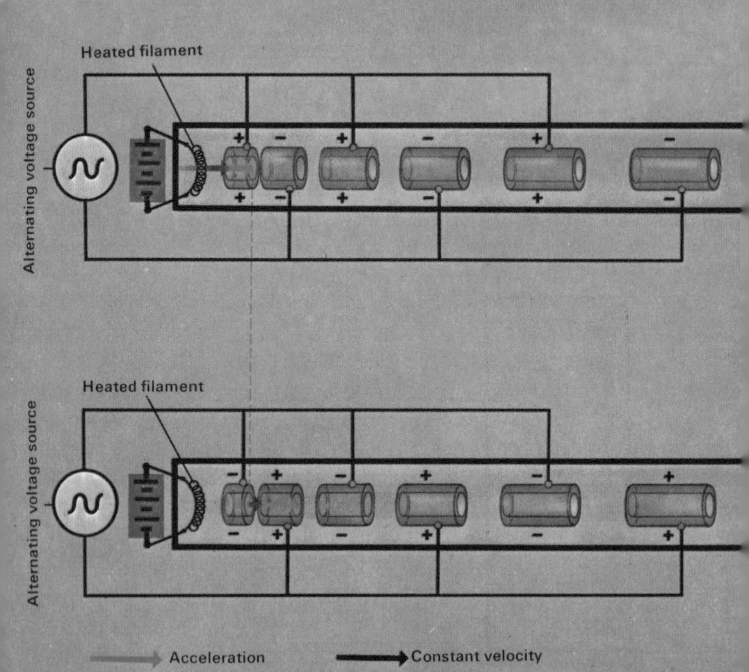

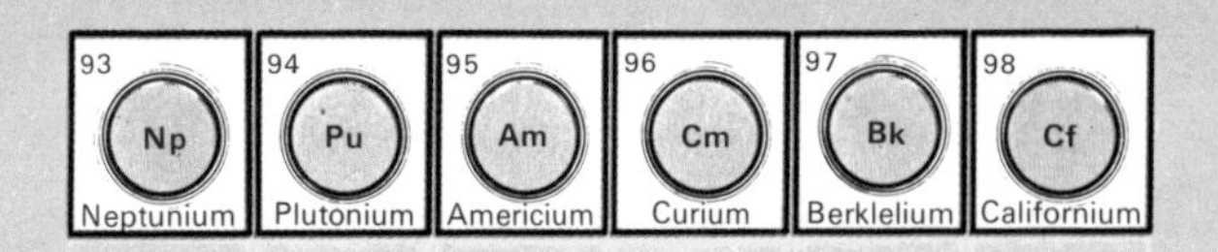

Table of transuranic elements

Beams of high-energy particles are required not only in the production of transuranic elements but also in the study of the elementary particles themselves. The kinetic energy of a charged particle is increased to these high values by accelerating it in an electric field inside a machine called an accelerator. One such device is the *linear accelerator,* in which particles – electrons or protons – travel in straight lines down an evacuated tube. There are various types of linear accelerators; one type consists essentially of a row of cylindrical coaxial electrodes separated by narrow gaps. Alternate electrodes are connected together to opposite potentials of a source of high-frequency alternating voltage, such as a magnetron. As the voltage alternates, the charge on the electrodes changes sign. Electrons, injected into the tube, will only be accelerated towards a positively charged electrode. Once inside the electrode, their velocity remains constant. The length between the gaps is arranged so that the electrons emerge from the electrode after one half of the voltage cycle has passed, and the charge on the electrodes has changed sign. They are therefore accelerated towards the next electrode, now positively charged. As the velocity of the electrons increases, the length must be increased accordingly, so that the electrons always take the same time to travel between

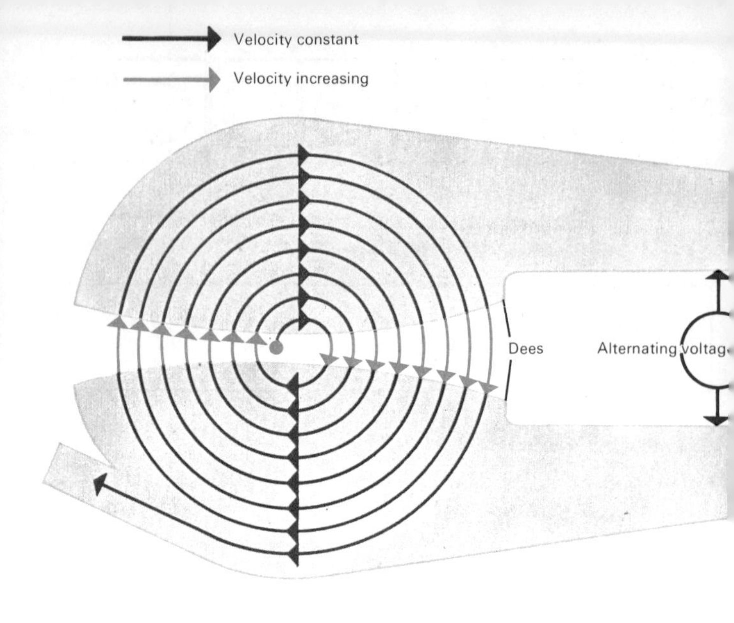

Diagram of cyclotron seen from above (*above*) and in cross-section (*below*)

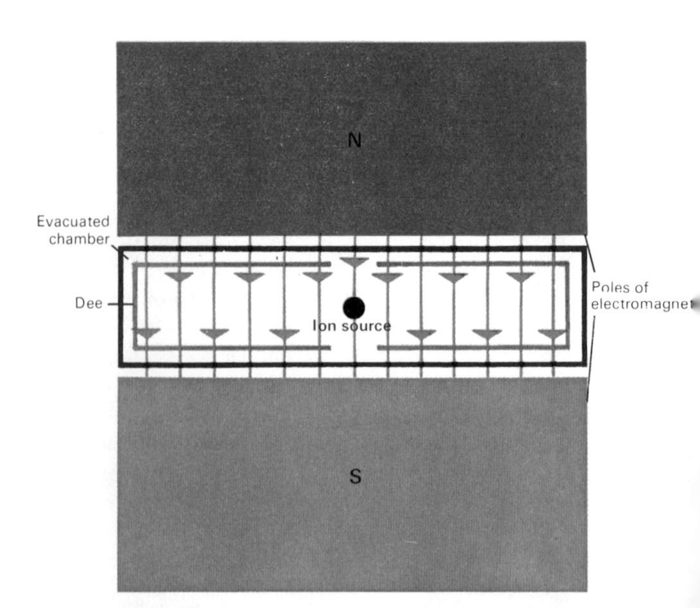

the gaps and consequently can keep in phase with the alternating voltage. This process is repeated down the length of the tube, the electrons gaining energy each time they are accelerated across a gap. The electrons emerging from the end of the tube have velocities approaching that of light and therefore have very high energy.

Cyclic accelerators are machines in which charged particles travel many times around a circular path; one such device is the *cyclotron*. It consists essentially of two evacuated, hollow, semicircular chambers, called *dees* (because of their shape), sandwiched between the poles of an electromagnet. The particle source lies in between the dees. An alternating voltage, applied across the dees, causes positively-charged particles, such as protons, to be accelerated towards the negatively-charged dee. Once inside the dee their velocity remains constant. Under the influence of the magnetic field, the protons follow a semicircular path, the radius of which depends on the velocity of the protons. The frequency of the alternating voltage and strength of the magnetic field (both constant) are selected so that the time of travel across the gap and around the semicircular path in the dee corresponds to one half of the voltage cycle. On emerging from one dee the protons are therefore accelerated towards the other dee, now negatively-charged, hence increasing their energy. The process is repeated again with the increased velocity causing the radius of the semicircular path to increase. The time of revolution depends only on the magnetic field and is independent of the velocity; the velocity increase is exactly compensated by the longer travel path. When the protons leave the cyclotron their energy is very high. The cyclotron is the simplest of the cyclic accelerators. Energies many thousands of times greater are obtained in more advanced machines.

Using high energy particles, research can be carried out into the nature and properties of elementary particles. It is possible to classify particles according to their properties such as charge and spin, into a table resembling the periodic table of elements. This method, called *unitary symmetry,* has been successful in predicting the existence of particles that have subsequently been detected experimentally.

The detection of charged particles

The reactions and decay of elementary particles are observed and studied in several ways. The *bubble chamber* is an instrument in which the tracks of an ionizing particle are made visible as a row of bubbles in a liquid. The liquid in the bubble chamber is maintained under pressure so that its boiling point can be raised; it can therefore be heated to a temperature slightly above its normal boiling point without boiling. About one hundredth of a second before a particle enters the chamber, the pressure is suddenly reduced. The particle causes ionization of the atoms of the liquid, resulting in a release of energy that causes rapid localized boiling along the particle's path. After about one thousandth of a second the bubbles formed are large enough to be photographed and a record of the particle's track and that of its decay products is obtained. The pressure is then increased again to prevent the bulk of the liquid boiling.

The energy of charged particles, such as α– or β– particles, or of X-rays, can be measured by an *ionization chamber*, which contains a thin wire anode and a coaxial cylindrical cathode between which a low voltage is applied. The passage of a particle causes ionization of the gas atoms in the chamber; positive ions are attracted towards the cathode. electrons towards the anode. An electric pulse is therefore produced which is proportional to the energy of the radiation. The *Geiger Counter* works on the same principle but a much higher voltage is used. The ions formed have sufficient energy to produce further ionization, resulting in an *avalanche* of ion pairs. The charge collected by the electrodes is proportional to the number of primary ion pairs and several thousand particles per second can be counted. The amplification is so great that a single β–particle can be detected.

Ionizing particles can also be detected and counted by a *scintillation counter*, consisting of a scintillation crystal and a *photomultiplier*. An incident particle will cause a scintillation crystal, such as sodium iodide, to emit a flash of light that is detected by the photomultiplier whose photosensitive surface emits electrons. The time between the passage of the particle and the production of the current is extremely short so that individual particles can be counted.

Tracks in a bubble chamber

Ionization chamber (*upper*) and scintillation counter (*lower*)

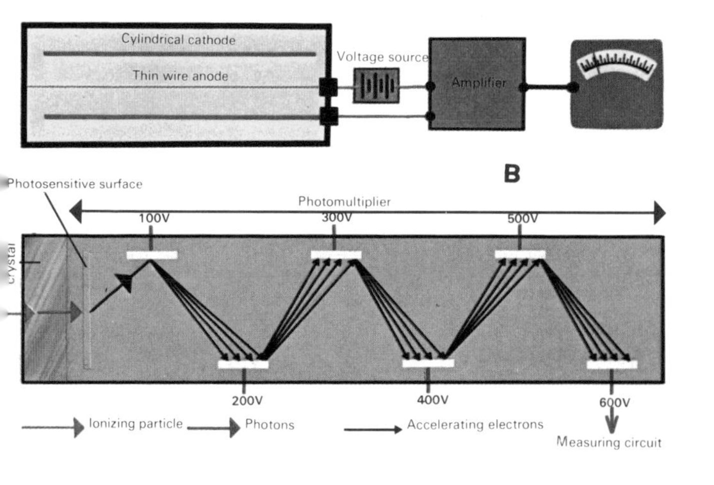

Nuclear fission

One of the two ways in which a nuclear reaction can produce energy is the fission of heavy atoms. In 1939 Otto Hahn discovered that neutron bombardment of substances containing uranium produced small quantities of lighter elements that had been formed by fission; it was subsequently found that each fission process liberated about 3×10^{-11} joules of energy. This in itself is not a large quantity of energy, but as 1 kg. of uranium contains about 2.5×10^{24} atoms, it follows that the complete fission of this quantity of uranium would produce about 7.5×10^{13} joules. For comparison, 1 kg of coal releases about 4×10^{7} joules on complete combustion. This means that about 1 ton of uranium will produce the same amount of energy as 2 000 000 tons of coal. Once it was realized that the atomic nucleus is such a rich source of energy intense efforts were made, especially during the last war, to find a method of releasing this energy.

Natural uranium consists of three isotopes: for every 100 000 atoms of the natural metal there are only 6 atoms of the isotope ^{234}U and 720 atoms of the isotope ^{235}U, all the remaining atoms being ^{238}U. When the ^{238}U nucleus is hit by a fast neutron (one travelling more than 17×10^{6} m/s) it undergoes fission and produces two lighter nuclei. However, when the ^{238}U nucleus is hit by a slower neutron it captures it, forming another isotope ^{239}U. This isotope is very unstable and decays into a new element, neptunium–239 (^{239}Np), by emitting a β–particle (electron); this element in turn decays into ^{239}Pu, again by emitting a β–particle.

The importance of ^{239}Pu and the rare uranium isotope ^{235}U is that they are both easily fissioned by neutrons of *all* speeds (they are *fissile*), and more important still, each fission is accompanied by the emission of either 1, 2, or 3 fresh neutrons. These fresh neutrons are able to produce their own fissions and thus a chain reaction builds up. As the fission process takes less than a microsecond, a few kilograms of ^{235}U or ^{239}Pu will disappear in one devastating explosion.

^{235}U and ^{239}Pu can be stored perfectly safely in small quantities (a few kilograms) as so many neutrons escape from the surface of the material that the chain reaction cannot be maintained. The smallest quantity in which a chain reaction

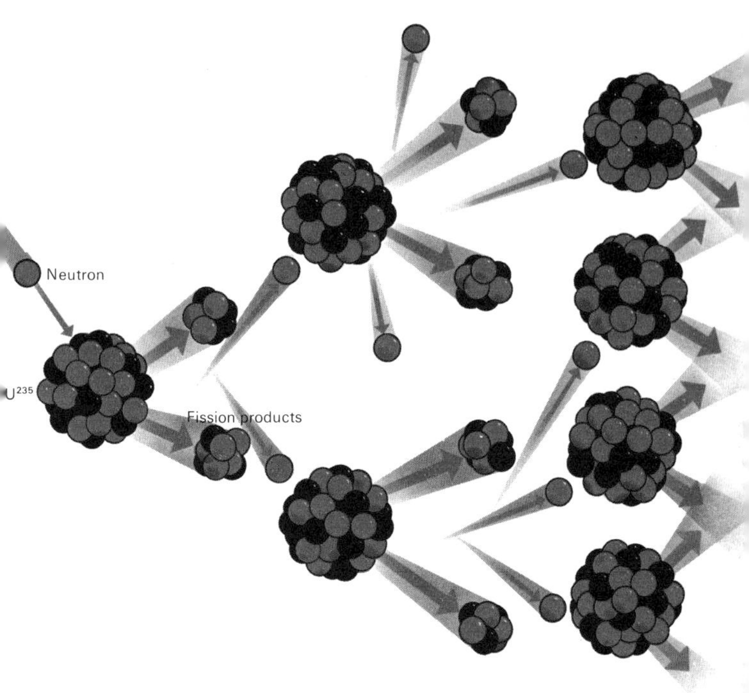

Chain reaction in ^{235}U

Formation of plutonium from ^{238}U

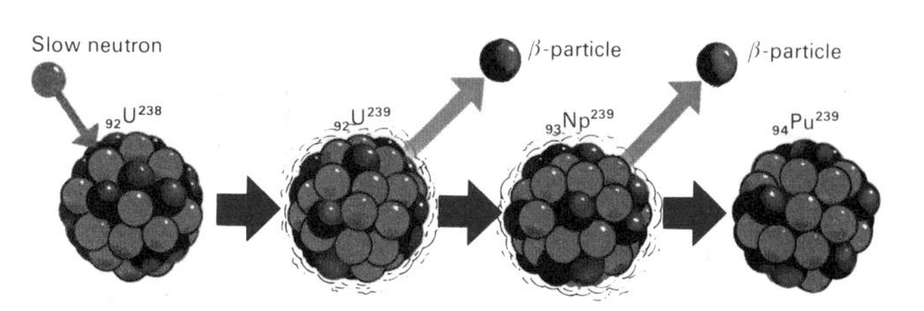

can be maintained is called the *critical mass*. If the critical mass is exceeded, the surface area per unit volume is decreased and enough neutrons remain within the material to keep the chain reaction going. An atom bomb consists of two subcritical masses of ^{235}U or ^{239}Pu and a mechanism for uniting them suddenly into a supercritical mass.

Controlled fission

In the atom bomb all the energy of the nuclear reaction is released in a fraction of a second – this type of uncontrolled chain reaction is therefore unsuitable for the peaceful uses of atomic energy. A nuclear reactor in a power station or ship is required to provide a steady and continuous supply of energy. There are two ways in which this can be done – one uses a *thermal reactor* and the other a *fast reactor*.

A thermal reactor uses a *moderator* to slow down the neutrons released in the fission reaction. Using natural uranium fuel, a neutron of any velocity will fission a ^{235}U atom that it happens to hit, but as only 72 atoms in every 10 000 are of this isotope most collisions will be with the plentiful ^{238}U isotope which absorbs fast neutrons. As most neutrons released by the fission process are fast, in natural uranium fuel a chain reaction cannot be sustained. However, if the neutrons are slowed down (becoming *thermal neutrons*),

Advanced gas-cooled reactor

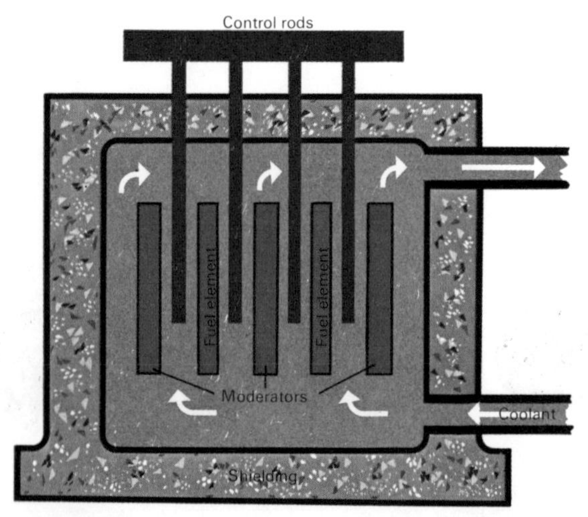

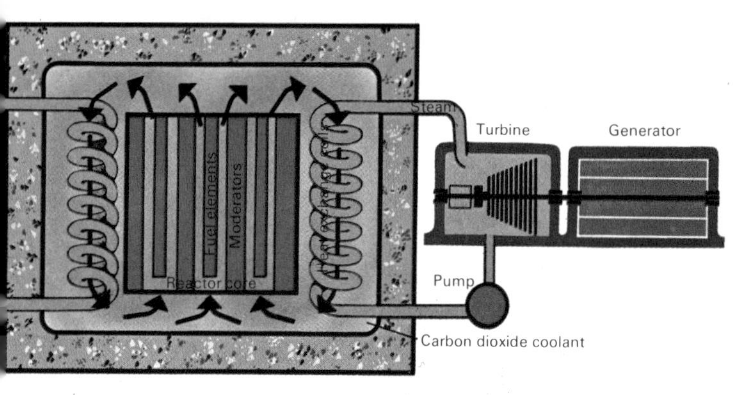

Thermal reactor

so that they have energies comparable to the thermal energy of vibration of the atoms amongst which they move, many more fissions of ^{235}U will occur because fewer of them will be absorbed by ^{238}U. This slowing down, or *thermalizing*, of neutrons is achieved by adding a substance called a moderator to the fuel or by surrounding the fuel elements with it. The moderator must consist of light atoms that will not capture neutrons but will reduce their energy to thermal levels after repeated collisions. Suitable substances are graphite (a pure form of carbon) and heavy water (water made from deuterium instead of hydrogen – D_2O).

In fast reactors no moderator is used but the frequency of collisions between neutrons and fissile atoms is increased by enriching the natural uranium fuel with ^{239}Pu atoms or additional ^{235}U atoms. The fast neutrons in this way build up a self-sustaining chain reaction, but not of course as fast as the chain reaction in the bomb. In these reactors the central core is usually surrounded by a blanket of natural uranium into which some of the neutrons are allowed to escape – under suitable conditions some of these neutrons will be captured by the ^{238}U atoms to form plutonium. As more plutonium can be produced than is required to enrich the fuel in the core, these are called *fast breeder reactors* (FBR).

In both thermal and fast reactors, the second to second

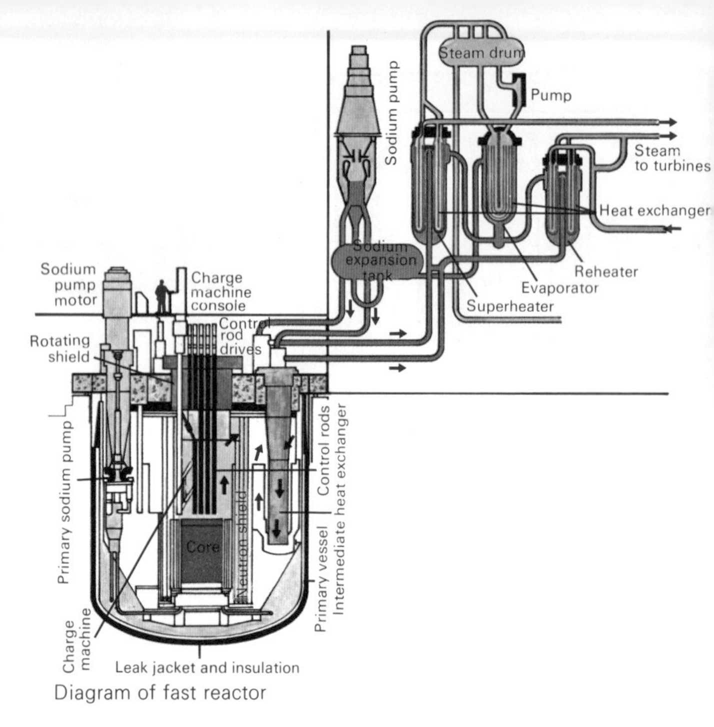

Diagram of fast reactor

Nuclear power station linked to desalination plant

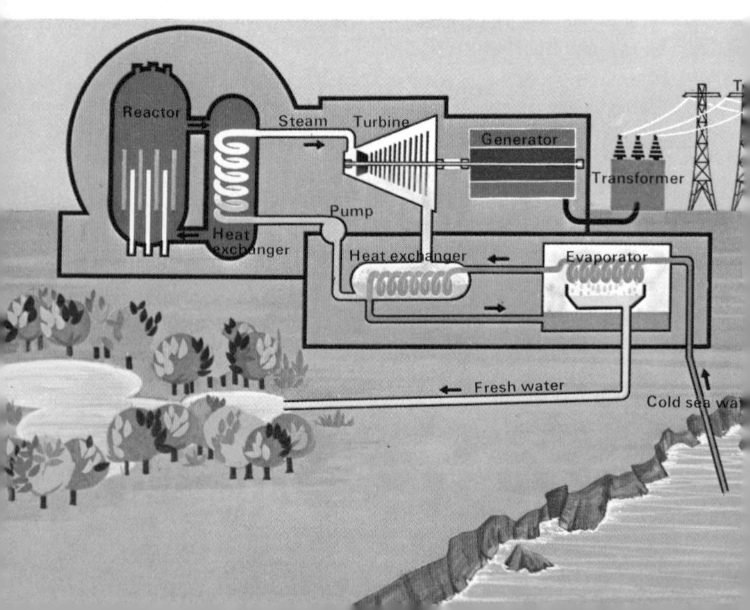

progress of the nuclear reaction is controlled by rods of neutron-absorbing materials that are lowered into the core. These *control rods* usually contain cadmium, hafnium, or boron and are raised and lowered automatically to maintain a steady reaction rate.

Most of the energy released by the nuclear reaction is given off as heat – the two nuclei into which the fissile nucleus splits (the *fission products*) have very high energies. They collide with the other fuel atoms and with the walls of the fuel elements increasing the energy of all these atoms and thus raising their temperature. This heat is extracted from the core by means of a *coolant* gas or liquid. The hot coolant is then used in a variety of ways to produce steam to drive a turbine and generator in the same way as a conventional power station.

Types of reactor

The first nuclear power station in the world was Britain's thermal reactor at Calder Hall (Cumberland) which was opened in 1956. Since them most of the major powers have built thermal reactors of one type or another. However, the next generation of nuclear reactors, for the 1980s, will be fast breeder reactors. Russia, Britain, France, and America all have plans to produce FBRs by the early 1980s (probably coming into commission in this order). The reason for the change to FBRs is that they can utilize 75 per cent of the uranium ore as it comes from the earth, compared to less than 1 per cent in thermal reactors. Moreover after the first fuelling, which takes about 3000 kg of plutonium per 1000 megawatts of electricity produced, the net input of fuel is a very small quantity of natural uranium. Fuel costs for fast reactors are expected to be half those for thermal reactors.

Over the past twenty years in Britain, the main line of reactor development has been in gas-cooled thermal reactors. The latest type, the advanced gas-cooled reactor (AGR), uses ceramic uranium dioxide fuel encased in stainless steel elements. These elements are channelled into stacks of graphite blocks which act as a moderator. The coolant, carbon dioxide gas, flows through the channels over the fuel elements and leaves the reactor core at a temperature of about 600°C. This hot gas passes through heat exchangers in which ordinary water is boiled to raise steam.

Shielding is necessary in all nuclear reactors to provide protection against neutron and gamma-radiation emitted during the decay of the fission products. Usually the shield consists of several feet of concrete surrounding an inner steel lining.

FBRs work at higher temperature than thermal reactors and the coolant used is usually liquid sodium, a metal with excellent thermal properties melting at 98°C and boiling at 880°C. Two sodium circuits are needed: in the first the sodium passes through narrow channels round the fuel elements, becoming radioactive. The sodium in this circuit is heated to a temperature of more than 500°C. Heat exchangers are then used to transfer some of this heat to another sodium circuit which requires a second heat exchanger to raise the steam to drive the turbines.

Nuclear submarine

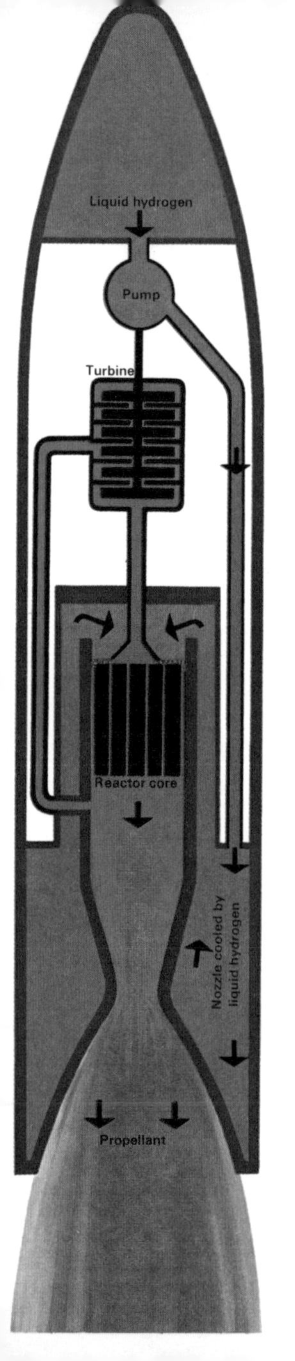

Other uses of nuclear reactors

Apart from the increasing use of nuclear reactors in power stations to produce electricity, more compact reactors have been developed for other uses – perhaps the most important being marine propulsion. Pressurized water is used as moderator and coolant.

The first nuclear submarine, the American *Nautilus,* was launched in 1954. The great advantage of nuclear power for submarines is that they can remain submerged for almost indefinite periods; an ordinary submarine has to surface frequently to charge its batteries by means of a diesel-driven generator requiring air. The first *Nautilus* reactor lasted for 62000 miles without refuelling, including an underwater crossing of the Arctic ice.

Nuclear ships have also been built by several countries; although they are not yet competitive with oil ships for general purposes, for special uses such as ice-breaking in polar waters they are ideal.

Nuclear powered rocket

Nuclear reactors are also beginning to feature in space programmes. In 1965 an American satellite was powered for 43 days by a fast nuclear reactor in which the coolant was a mixture of sodium and potassium. Heat was converted to electricity by passing the coolant over thermocouples. America also has plans for a nuclear rocket. The present design is a graphite moderated thermal reactor using hydrogen as the coolant. The hydrogen is heated to 2000°C in the reactor and is then used as the propellant. In order to withstand this temperature the nozzle is cooled by liquid hydrogen, in which form the hydrogen is stored in the rocket.

Nuclear fusion

The binding energy diagram on page 122 shows that energy is released by the fusion of light atoms as well as the fission of heavy atoms. Fusion reactions produce the energy in the sun and stars and therefore our life on earth depends on them. Stars, like the sun, probably began their existance as clouds of hydrogen – protons, neutrons, and electrons. The gravitational forces between the atoms would tend to make such a gas denser in the interior than on the outside. The collisions between atoms in the centre of the gas would therefore be very frequent and very energetic: the temperature and pressure in the interior would be very high indeed. As the temperature increases and the collisions become more vigorous, electrons are dislodged from their orbits – the atoms become ionized. An ionized gas like this is sometimes referred to as the fourth state of matter – it is called a *plasma*. Within this plasma, if the temperature is high enough, protons combine with neutrons to form deuterium nuclei and the deuterium nuclei combine to form tritium and helium nuclei. These fusion reactions will only occur above a certain temperature, called the ignition temperature. Below this temperature too much energy is lost by *bremsstrahlung radiation* – the direct conversion of kinetic energy into X-rays as a result of a near collision between an electron and a proton or a nucleus.

The energy released in these fusion reactions is around 13×10^{-13} joules. As 1 kg of deuterium contains 3×10^{26} nuclei, the energy released by the conversion of this quantity of deuterium to helium would be 3×10^{14} joules, or about six times more than is released by the fission of 1 kg of uranium. Fusion is thus a potential source of enormous quantities of energy; moreover, it depends on a plentiful fuel

Fusion reactions of hydrogen (*upper*) and lithium (*lower*) isotopes

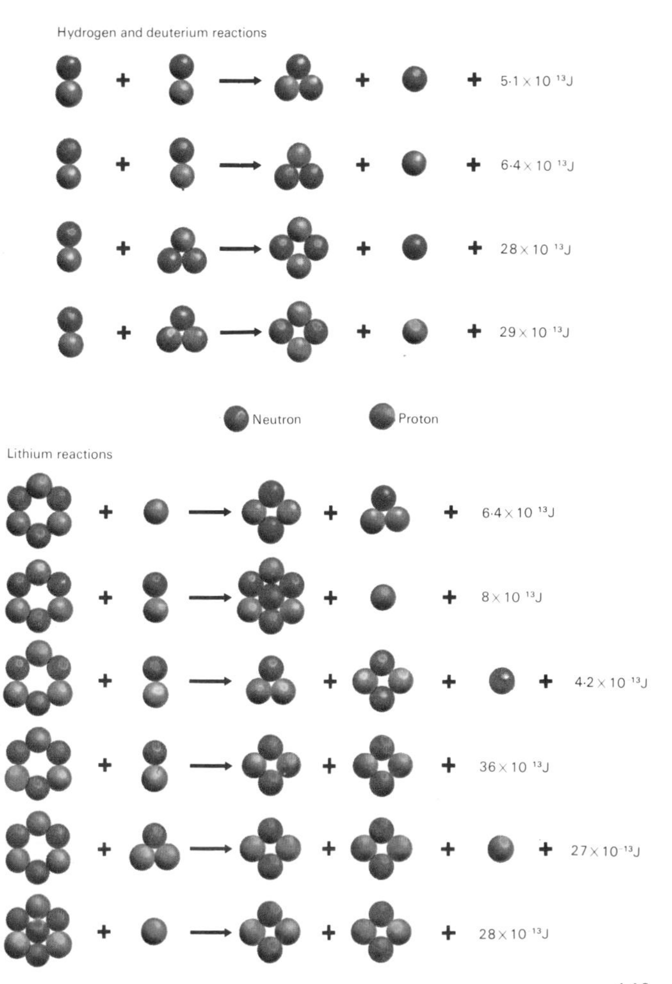

– the sea is full of hydrogen and 15 out of every 100 000 hydrogen atoms are deuterium atoms. However, the problem is to create the extremely high ignition temperature (about 50 000 000°C) of the deuterium reaction. The only known source of such a temperature is the atom (fission) bomb. In 1952 the first fusion reaction on earth was created by encasing a fission bomb with a layer of hydrogenous material – this was the *hydrogen bomb* – and it had a much greater destructive power than the fission bomb. The fission bomb dropped on Hiroshima had a destructive power equivalent to 20 kilotons of TNT – a fission-fusion bomb is equivalent to 20 megatons of TNT. Lithium deuteride is the fusion material.

Controlled fusion

The problems of harnessing fusion reactions (or *thermonuclear reactions* as they are sometimes called) for peaceful uses have not yet been solved, although intense research is going on in many places in the world. There are three main problems: creating the ignition temperature, containing the reaction in a vessel (all known substances vaporize at a few thousand degrees Celsius), and recovering the energy of the reaction.

In 1963 the ignition temperature of the deuterium-tritium reaction was reached in an experimental device for a very short period (3 μs). This was achieved by passing an enormously strong pulse of electric current through a low pressure mixture of the gases. This pulse not only ionized the gases, it also created a very powerful magnetic field which prevented the particles in the plasma from touching the walls of the tube. It does this because the magnetic field makes the charged particles move in helical tracks around the lines of force – the stronger the field the narrower the helices. Thus the particles only move up and down the tube, never across it. This contraction of the plasma by a magnetic field, called the *pinch effect,* partially solves the containment problem. Unfortunately, however, the confined plasma is not stable for any length of time as it develops kinks which could

(*Upper*) Possible scheme for a fusion reactor
(*Middle*) The three parameters of a fusion reaction
(*Lower*) Zeta pinch (*left*) and theta pinch (*right*) plasma confinement

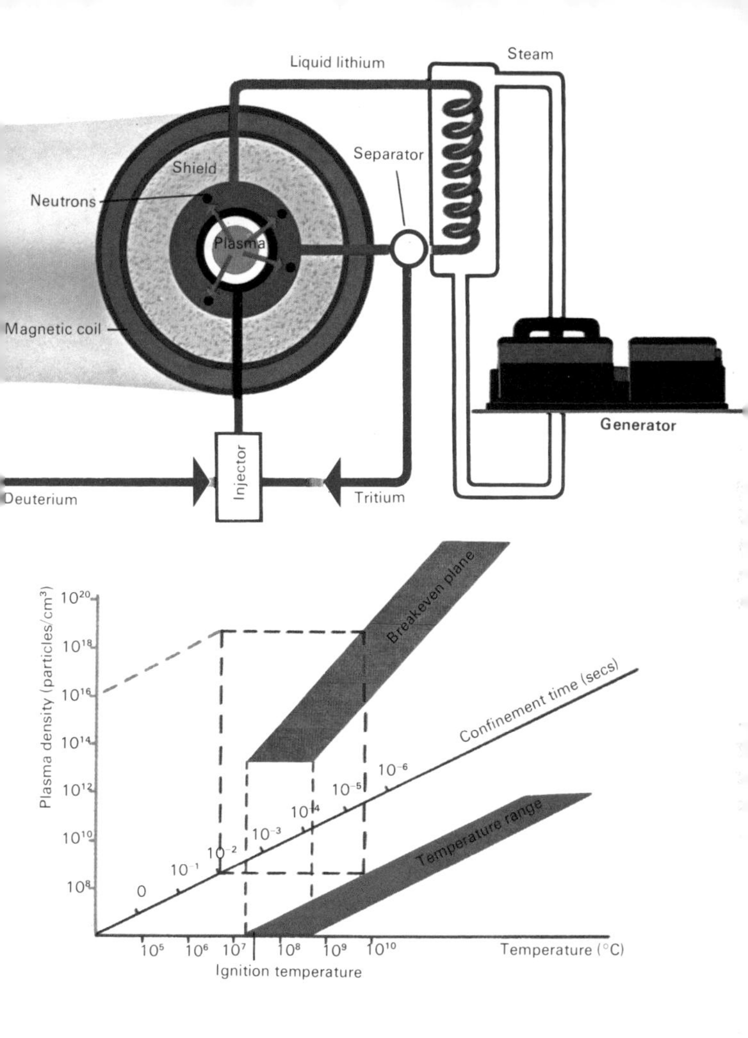

Liquid lithium
Steam
Shield
Separator
Neutrons
Plasma
Magnetic coil
Generator
Injector
Deuterium
Tritium
Breakeven plane
Confinement time (secs)
Plasma density (particles/cm³)
Temperature range
Temperature (°C)
Ignition temperature

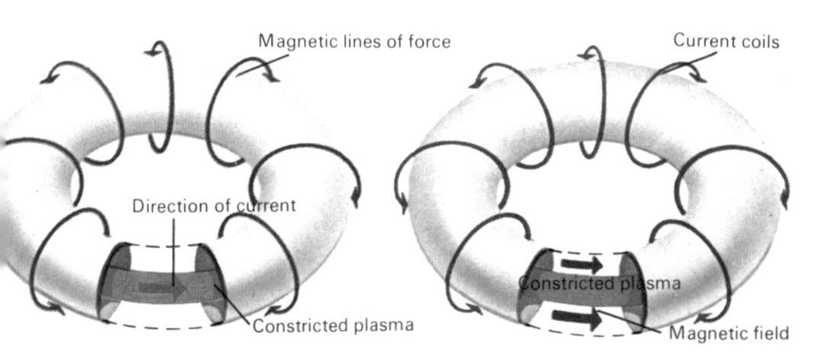

Magnetic lines of force
Direction of current
Constricted plasma
Current coils
Constricted plasma
Magnetic field

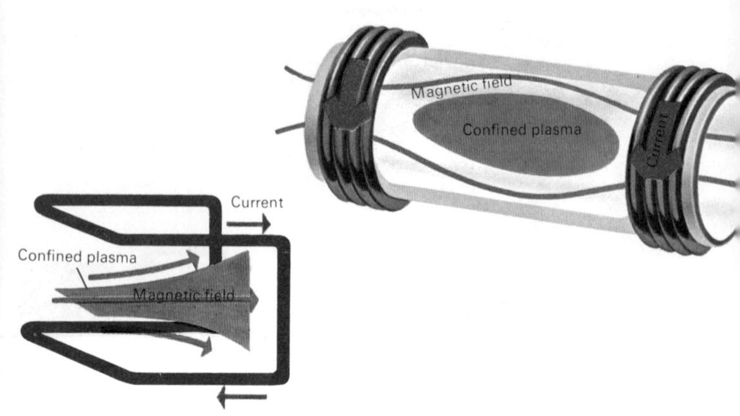

Plasma confinement in baseball bar (*left*) and linear magnetic bottle stopped by magnetic mirrors (*right*)

bring it into contact with the walls of the tube. Much of today's fusion research is concerned with designing systems in which a magnetically confined plasma remains stable long enough for energy to be extracted from the reaction.

The pulse that creates the plasma and raises it to the ignition temperature clearly requires a considerable amount of energy. In order for the reaction to provide a surplus of energy the plasma has to be confined for a certain minimum time, at any given temperature (above the ignition temperature) and any given density. J. D. Lawson found that for a deuterium-tritium plasma, the confinement time multiplied by the density (seconds $\times$ particles per cm^3) had to be at least 10^{14} to reach a breakeven level where exactly enough thermonuclear energy is produced to provide the pulse energy. This *Lawson criterion* is illustrated as a plane over a range of temperatures from the ignition temperature to 5×10^8 °C. No device has yet managed to reach this plane, although in some devices two of the three necessary values (temperature, density, and confinement time) have been reached.

Greater stability in these devices can be achieved by incorporating extra current-carrying structures into the tube. The first structures used were straight rods called Ioffe bars after the Russian physicist M. S. Ioffe. Other structures create a magnetic well by using baseball bars (so called because they resemble the seam on a baseball).

These aspects of research into the design of future fusion reactors are being carried on concurrently with investigations into the best method of extracting the energy. The simplest type of reactor, using a deuterium-tritium reaction, releases 80 per cent of its energy in the form of energetic neutrons. In this type of reactor the neutron energy could be absorbed in a liquid lithium coolant surrounding the reactor tube. One advantage of this method is that the neutrons could also be used for making fissile fuel for fission reactors from ^{238}U.

In fusion reactions that release most of their energy in the form of charged particles, such as the deuterium-deuterium reaction, the kinetic energy of these particles could be converted directly into electrical energy. This could be done by collecting the particles on a series of positively or negatively charged electrodes. By applying suitable voltages to these electrodes the conversion efficiency could be as high as 90 per cent.

Fusion reactor used for generating electricity directly by collection of charged particles

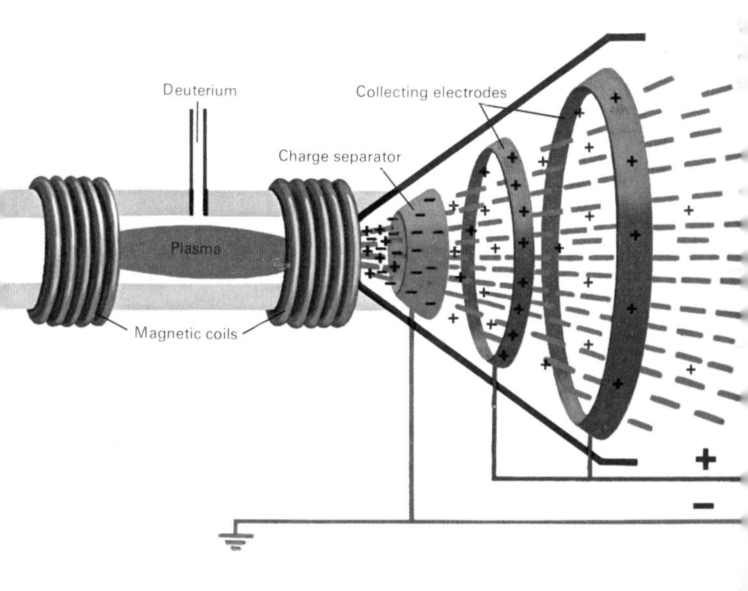

RELATIVITY

Einstein proposed his celebrated theory of relativity in 1905 in order to explain the result of an experiment performed in 1887 by Michelson and Morley. The purpose of this experiment was to discover whether the velocity of light depended on its direction of motion relative to the earth: they compared the velocity in the direction of the earth's rotation with the velocity at right angles to the rotation. As they found no difference in the velocity in these two directions they were forced to conclude that the velocity of light does not depend on the velocity of its source or, putting it another way, that the velocity of light is absolute and is not relative to the velocity of the observer. This was a very startling conclusion.

The classical law for relating velocities is a simple addition or subtraction law, $v_{AB}=v_A \pm v_B$, where v_{AB} is the velocity of A relative to B when v_A is the velocity of A and v_B is the velocity of B, the sign depending on whether or not the velocities are in the same direction. However, if an astronaut in a spacecraft flying at 90 per cent of the velocity of light ($0.9c$ where c is the velocity of light) measures the velocity of a beam of light coming towards him from a spacecraft travelling in the opposite direction, he will find that it is not $c+0.9c=1.9c$, but c. The law for relating velocities proposed by Einstein is not a simple addition v_A+v_B but $v_{AB}=$

The Michelson-Morley experiment attempted to show a difference in the velocity of light along the paths S–B, A–C.

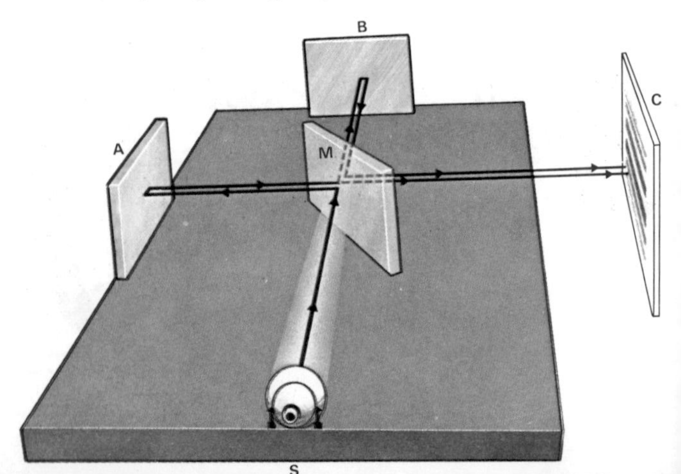

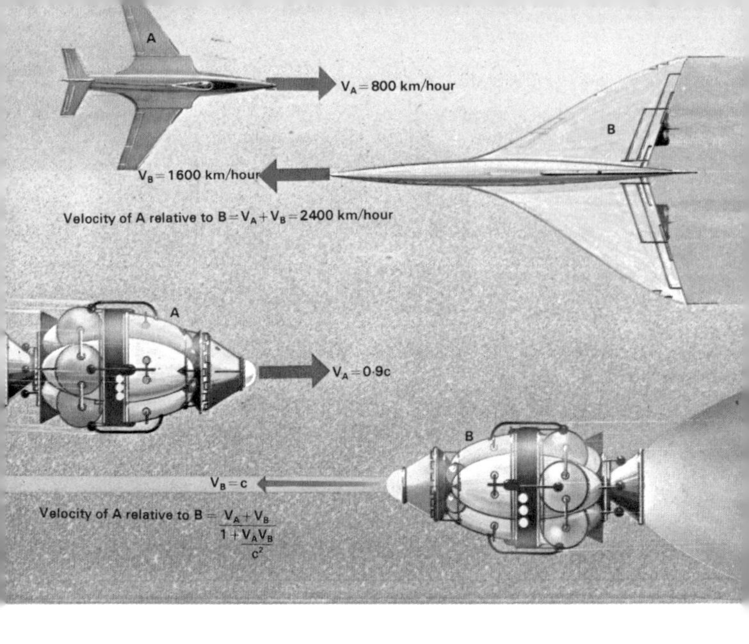

The classical law for relating velocities does not apply to bodies travelling at velocities approaching that of light.

$(v_A + v_B)/(1 + v_A v_B/c^2)$. Substituting c for v_B in this equation gives $v_{AB} = c$. This is a general law for relating all velocities and it explains the result of the Michelson-Morley experiment. However, if v_A and v_B are small compared to c the equation reduces to the simple $v_A + v_B$. As c is very large, 3×10^8 m/s, even if v_A and v_B are 2000000 km/hour there is less than 1 per cent difference between the two ways of calculating v_{AB}.

This is the basis of the *special theory of relativity* – it is called the special theory because it applies only to bodies travelling at constant velocities; the *general theory* applies to accelerating bodies. The special theory has some very far-reaching consequences.

First, it can be shown from the relativity of velocities that if the length of B's spacecraft is L_B at rest, when it is travelling at v_{AB} relative to A it will appear to the astronaut in A to have a length $L_{AB} = L_B\sqrt{(1 - v_{AB}/c^2)}$. This contraction in length, called the *Lorentz-Fitzgerald contraction* after the two scientists who independently proposed it, is negligible at low velocities. At 800 km/hour the contraction is about 10^{-12} cm, but at $0.9c$ it halves the length. Neither of the two observers in the spacecraft will notice any difference to their own crafts, of course, as they are travelling with them; the effect would only be noticeable to observers in relative motion.

A further consequence of special relativity is the increase of mass with velocity, again expressed by a similar formula: $m_v = m_0/\sqrt{(1 - v^2/c^2)}$, where m_v is the mass at velocity v and m_0 is the *rest mass*. This equation also shows that at normal velocities the increase of mass is negligible, but at $0.99c$ the mass of a body is seven times its rest mass.

An even more profound effect of this mass increase is the relation between mass and energy that results directly from it. Einstein showed that the energy, E, of a quantity of mass, m, is given by $E = mc^2$. Thus mass is a form of energy and when it is destroyed it reverts to energy.

Yet another consequence of special relativity is that if two

Astronomical proof of general relativity. Bending of light in sun's gravitational field gives the star – visible during total eclipse of the sun – an apparent displacement.

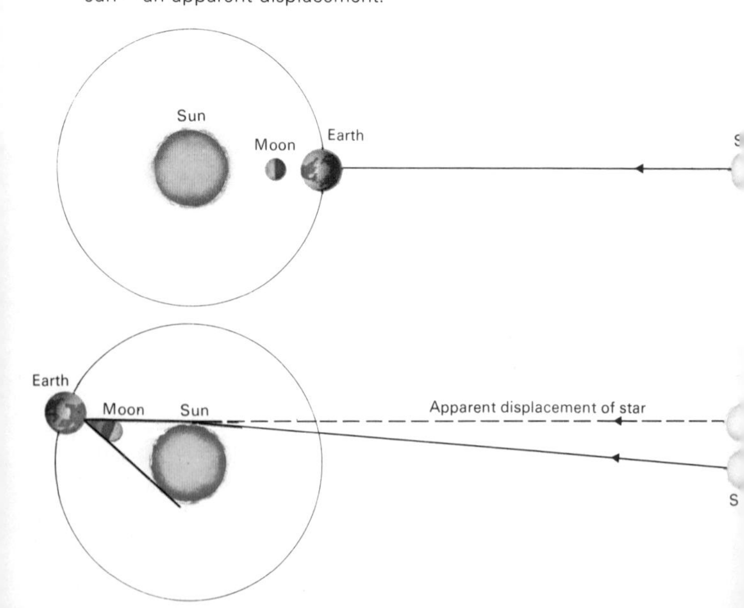

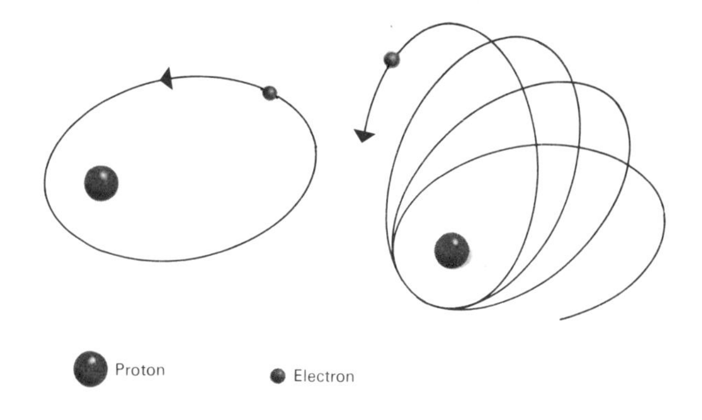

Elliptical orbit of electron assuming stationary mass (*left*)
Precession of orbit for mass of moving electron (*right*)

observers are moving at constant velocity relative to each other, it will appear to each that the other's clocks have been slowed down. This affects the concept of simultaneous events. If an event on the sun, say, occurs at a certain instant t, we do not know of it on earth until $t+8$ minutes because it takes 8 minutes for light from the sun to reach us. To an observer on the distant planet Jupiter the event would not be known until $t+43$ minutes. To observers on earth and observers on Jupiter the event would not appear simultaneous. To pinpoint an event, therefore, Einstein realized that both times and distances have to be considered together. He therefore suggested that space be treated as having four dimensions – three of space and one of time forming a four-dimensional continuum.

When in 1916 Einstein published his *general theory of relativity* he based it on a four dimensional continuum and found that Euclidean geometry was no longer valid. A Euclidean straight line follows the curvature of the earth, which is satisfactory for earth measurements, but in space a straight line is determined by the path of light rays. As Einstein predicted that light rays would be attracted to massive bodies by gravitational forces, he described space as being 'curved' to an extent that depends on the masses it contains. This bending of light rays by gravitational attraction has since been confirmed by astronomical observations.

UNRESOLVED THEORIES OF PHYSICS AND COSMOLOGY

This book gives some idea of what physics is about and outlines some contemporary theories in physics. But, like every other science, physics is continually expanding – it does not remain static. New observations frequently require new explanations and, what is more, they may require the overthrow of theories that have seemed unassailable. Physicists have to keep an open mind, being prepared to abandon long-standing theories if they no longer satisfactorily explain new observations.

As well as being prepared to change existing theories, physicists have to seek new theories for problems that are still unsolved. Perhaps the most puzzling problem that still confronts us is the relationship between the four types of forces that are known to exist in the universe.

Electromagnetic and gravitational interactions are normally

Heisenberg (*left*) and Planck (*right*)

expressed in a similar form. The coulomb (electrostatic) force between charged particles is proportional to q_1q_2/d_2, the magnetic force between poles of strength M_1 and M_2 is proportional to M_1M_2/d^2, and the gravitational force between two masses is proportional to m_1m_2/d^2. The form of these equations suggests that there ought to be some way of expressing all of them in one set of equations, even though the gravitational force differs from the other two in that it is always a force of attraction. However, a *unified field theory,* which Einstein sought for many years, has never been found.

It may be that a unified theory will have to abandon the field concept, so that all types of force will be expressed in terms of an exchange of particles between interacting bodies. The strong meson exchange force between nucleons has already been mentioned in these terms. The electrostatic forces between charged bodies can be interpreted as an exchange of photons between charges. However, as these photons would exist for an extremely short time and cannot be detected, they are called *virtual photons.* Likewise, the exchange of a postulated particle, the *graviton,* could explain gravitational forces between massive bodies. The nuclear forces are 100 times stronger than the electromagnetic forces, 10^{12} times stronger than the weak interactions that occur during β–decay, and 10^{39} times stronger than the gravitational forces. One day, this wide range of magnitudes may be expressed in one set of equations. Such a theory should also be able to accommodate both wave and particle aspects of these interactions.

Protons, neutrons and electrons tend to be thought of now in much the same way as chemists at the end of the last century thought of atoms. But it is possible that some of these particles themselves have a structure. One currently popular theory, which so far lacks experimental proof, is that all baryons and mesons are made up of groups of more fundamental particles called *quarks*.

Perhaps the greatest success of modern physics has been in understanding the relationship between mass and energy. We now know that they are interchangeable aspects of the same thing: mass can be converted into energy and energy can be turned into mass. *Pair production* is an example of the creation of matter: undetectable virtual photons and virtual electrons explain the transfer of energy and momentum between an interacting photon and proton and the interaction results in the production of a real electron-positron pair.

The theory of elementary particles, with these real and virtual states, is clearly confused at present. It needs a unifying central concept to bring together all the threads that now have to be treated separately.

Another quite different aspect of our ignorance concerns the origin of the matter and energy in the universe. There are two theories: in one, called the *steady state theory,* the universe is assumed to have always existed. Although it is usual to think of things having a beginning and an end, there is no logical reason for assuming that this must be so. The universe may have always existed and always continue to exist.

The conflicting theory assumes that the universe did have a beginning, when all the mass and energy of which it consists were concentrated into one *superdense* agglomeration. At a certain point in time (in some unexplained way and for some unexplained reason) this agglomeration exploded, lumps of matter being flung into space like fragments from an exploding bomb. This is called the superdense or *big bang* theory. What evidence is there to support either of these theories?

The red shift of the galaxies provides some evidence. The further away a galaxy is from ours, the greater is its red shift. Interpreting red shifts as a Doppler effect indicating recession,

Pair production – an energetic photon interacts with a proton to form an electron and a positron.

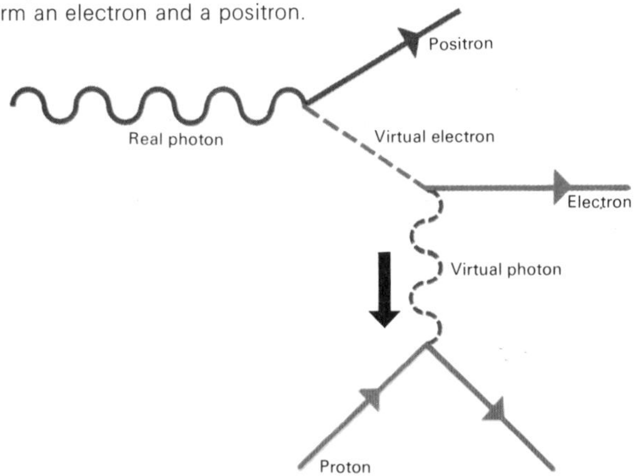

Positron
Real photon
Virtual electron
Electron
Virtual photon
Proton

the red shift means that the galaxies are flying apart and that the further away a galaxy is from us the faster it is receding. This observation can be taken to support the big bang theory – in which case it is possible to estimate the time that has elapsed since the explosion. This period is known as the Hubble constant and it is estimated as being between 10 and 40×10^9 years.

Supporters of the steady state theory explain the *expansion of the universe* by assuming that matter, in the form of hydrogen, is being continuously created and that the universe is expanding to maintain a uniform density. The rate of production of hydrogen in order to maintain this uniform density would be only one atom per cubic metre in about 300 000 years. This rate is far too low to be observed and therefore evidence has to be sought elsewhere.

One vital piece of information would be the actual density of matter throughout the universe. If it were fairly uniform throughout space this would be considered support for the steady state theory. But if it is denser in the parts of the universe furthest away from us then the big bang theory might seem more likely. This is because the universe is so immense that light takes an enormous time to travel across it. A common unit for measuring distances in space is the *light year* – the distance travelled by light in one year (about 10^{16} metres). Light reaching us now from stars a million light years away gives us information about the state of the universe – such as

Hypothetical combinations of quarks to form neutrons and protons

q_1 charge $\frac{2}{3}e$ mass m

q_2 charge $-\frac{1}{3}e$ mass $m + m'$

q_3 charge $-\frac{1}{3}e$ mass $m + m''$

Neutron
charge
$-2/3e + 2(1/3e) = 0$

Proton
charge
$2(-2/3e) + (1/3e) = -e$

its density – a million years ago. Light from stars much further away in space brings information of events even closer to the time of the big bang.

The evidence, such as it is, seems to indicate that the universe used to be denser than it now is. But this does not constitute proof of the big bang theory – the evidence is not yet convincing enough; moreover, it has recently been suggested that the universe may oscillate between expansion and contraction – the present phase being one of expansion. It may be that we shall know the answer to these fascinating questions in the not too distant future.

APPENDIX

Although the object of this book is primarily to *describe* some of the physical aspects of science, rather than to give detailed methods of calculation, it has been necessary in some cases to give magnitudes of physical quantities. These physical quantities are expressed in the units now used by physicists all over the world – the SI (Système International) Units. Many of these will already be familiar (metre, second, volt, etc.), a few of the less familiar ones (newton, joule, etc.) are defined when they are used.

To express very large and very small numbers and quantities, a set of agreed prefixes, abbreviations, and symbols are used with SI units. These are given in the following table.

MAGNITUDE	ABBREVIATION	PREFIX	SYMBOL
Thousand million	10^9	giga-	G
Million	10^6	mega-	M
Thousand	10^3	kilo-	k
Thousandth	10^{-3}	milli-	m
Millionth	10^{-6}	micro-	μ
Thousand millionth	10^{-9}	nano-	n

For example, 1 nm = 1 nanometre = 10^{-9} metre = 0.000 000 001 metre

1 MV = 1 megavolt = 10^6 volt = 1 000 000 volts.

With SI Units it is customary to omit the comma after groups of three digits and to leave a space instead.

INDEX

Page numbers in bold type refer to illustrations

SOME OTHER TITLES IN THIS SERIES

Arts
Domestic Animals and Pets
Domestic Science
Gardening
General Information
History and Mythology
Natural History
Popular Science

Arts
Antique Furniture/Architecture/Clocks and Watches/Glass for Collectors/Jewellery/Musical Instruments/Porcelain/Pottery/Victoriana

Domestic Animals and Pets
Budgerigars/Cats/Dog Care/Dogs/Horses and Ponies/Pet Birds/Pets for Children/Tropical Freshwater Aquaria/Tropical Marine Aquaria

Domestic Science
Flower Arranging

Gardening
Chrysanthemums/Garden Flowers/Garden Shrubs/House Plants/Plants for Small Gardens/Roses

General Information
Aircraft/Arms and Armour/Coins and Medals/Flags/ Fortune Telling/Freshwater Fishing/Guns/Military Uniforms/Motor Boats and Boating/National Costumes of the world/ Orders and Decorations/Rockets and Missiles/ Sailing/Sailing Ships and Sailing Craft/Sea Fishing/Trains/Veteran and Vintage Cars/Warships

History and Mythology
Age of Shakespeare/Archaeology/Discovery of: Africa/ The American West/Australia/Japan/North America/South America/Great Land Battles/Great Naval Battles/Myths and Legends of: Africa/Ancient Egypt/Ancient Greece/Ancient Rome/India/The South Seas/Witchcraft and Black Magic

Natural History
The Animal Kingdom/Animals of Australia and New Zealand/Animals of Southern Asia/Bird Behaviour/Birds of Prey/Butterflies/Evolution of Life/Fishes of the world/ Fossil Man/A Guide to the Seashore/Life in the Sea/Mammals of the world/Monkeys and Apes/Natural History Collecting/The Plant Kingdom/Prehistoric Animals/Seabirds/Seashells/Snakes of the world/Trees of the world/Tropical Birds/Wild Cats

Popular Science
Astronomy/Atomic Energy/Chemistry/Computers at Work/ The Earth/Electricity/Electronics/Exploring the Planets/Heredity
The Human Body/Mathematics/Microscopes and Microscopic Life/Physics/Undersea Exploration/The Weather Guide